SpringerBriefs on Pioneers in Science and Practice

Volume 30

Series editor

Hans Günter Brauch, Mosbach, Germany

More information about this series at http://www.springer.com/series/10970
http://www.afes-press-books.de/html/SpringerBriefs_PSP.htm

Peter T. Coleman · Morton Deutsch

Morton Deutsch: A Pioneer in Developing Peace Psychology

Springer

The Morton Deutsch International Center for Cooperation and Conflict Resolution
Teachers College | Columbia University

Peter T. Coleman
The Morton Deutsch International Center
for Cooperation and Conflict Resolution,
Teachers College
Columbia University
New York, NY
USA

Morton Deutsch
The Morton Deutsch International Center
for Cooperation and Conflict Resolution,
Teachers College
Columbia University
New York, NY
USA

Acknowledgment: The cover photograph as well as all other photos in this volume were taken from the personal photo collection of the author who also granted the permission on their publication in this volume. A book website with additional information on Morton Deutsch, including videos and his major book covers is at: http://afes-press-books.de/html/SpringerBriefs_PSP_MortonDeutsch.htm

ISSN 2194-3125 ISSN 2194-3133 (electronic)
SpringerBriefs on Pioneers in Science and Practice
ISBN 978-3-319-15439-8 ISBN 978-3-319-15440-4 (eBook)
DOI 10.1007/978-3-319-15440-4

Library of Congress Control Number: 2014960344

Springer Cham Heidelberg New York Dordrecht London

Copyediting: PD Dr. Hans Günter Brauch, AFES-PRESS e.V., Mosbach, Germany

Project coordination: Elizabeth Hernandez, Project Coordinator, Morton Deutsch International Center for Cooperation and Conflict Resolution (MD-ICCCR), Teachers College, Columbia University, New York, USA

Proofreading: Ariel Bernstein, Tiffany Brown, Ljubica Chatman, Christine T. Chung, Ginevra Drinka, Jen Hull, Leslie Migliaccio, Kristen Rucki, Ling Shen, all Columbia University, New York, USA

Printed on acid-free paper

Springer International Publishing AG Switzerland is part of Springer Science+Business Media (www.springer.com)

To our great grandchildren,
may they live in a world
that is congenial, sustainable,
and free of war and destructive conflicts.

Morton Deutsch at his summer home in East Hampton, 1985. *Source* From the author's personal photo collection

Preface

Morton Deutsch, one of our great pioneers of peace, long believed in the power of ideas to rectify social problems and realize global well-being. Nurtured in the intellectual atmosphere of Karl Marx, Sigmund Freud, Charles Darwin, and Albert Einstein, he witnessed that power in action. Educated and trained to spar, question, and debate in the halls of the City University of New York, he honed his considerable intellect and skills as a social critic, theorist, and researcher. So as a young man returning from war, as the world was still reeling from the atrocities of the Nazi death camps and the annihilation of Hiroshima and Nagasaki, he turned his mind to the grandiose task of addressing social ills, big and small. From marital strife to global thermonuclear warfare, from ethnic slights between peers to race riots in the streets, and from acts of "civilized oppression" in the home to state-sanctioned policies of apartheid, Deutsch devoted his life to addressing social problems through the power of ideas.

This book celebrates the ideas of Morton Deutsch. It presents several of Deutsch's most seminal theoretical papers, all published previously in different journals and texts, but brought together here to highlight their originality, importance, and relevance for the world today. They cover much ground, from his early work on different types of interdependence in groups, conflict resolution, and social justice, to his later thinking on interrupting oppression and sustaining justice. They culminate in to two of his most ambitious intellectual undertakings: a general theoretical model of the psychosocial dynamics between people and different types of social situations, and his prophetic vision of the processes and institutions necessary for a more peaceful and prosperous world.

Embedded in each of these groundbreaking papers are the ideas, the distinctions, variables, relationships, dynamics, and outcomes that Deutsch developed through his reflection, research, and practice (He was an experimental scientist and a practicing psychoanalyst), which he delivered to us as a fully-developed theory. These ideas are essential. Although the theory has been defined as "An arbitrary structure that we impose on chaos to make it meaningful and predictable"

(John Whiting), some structures are much less arbitrary than others. Trained by Kurt Lewin to focus on "the essence of the phenomenon," Deutsch's focus in his theoretical work was basic and laser-like. Each insight integrated the work of others but then launched it to another level—to models of tremendous precision, resonance, and impact.

The focus of these ideas include: cooperation–competition in groups, the conditions that lead to constructive versus destructive conflict, Deutsch's Crude Law of Social Relations, the role of equity, equality, and need in distributive justice, the conditions for awakening a sense of injustice, a framework for interrupting injustice and oppression, the fundamental dimensions of social relations, and the dynamics between psychological orientations, social relations, and fit. This set of ideas and models are merely illustrative of the breadth and depth of Deutsch's theorizing and the implications of his ideas for addressing social problems today.

Through decades of extraordinary work, Morton Deutsch became an internationally renowned social psychologist, widely honored for his scientific and practical contributions and beloved by his students. He was one of the most important pioneers in the development of modern social psychology. He lead by example and consequently became a great leader of leaders, including Jeff Rubin, Roy Lewicki, David Johnson, Michelle Fine, Harvey and Madelaine Hornstein, Barbara Buncker, Susan Opotow, Eric Marcus, Ken Sole, Kenneth Kressel, and Adrienne Asch, to name only a few.

This book celebrates the life's work of Morton Deutsch, one of the world's most influential peace psychologists. A hard-nosed scientist of the Lewinian tradition, Deutsch invested his long and prestigious career in the development of a rigorous empirical approach to the study of cooperation, constructive conflict resolution, and social justice—the basic building blocks of sustainable peace.

New York, NY, USA, December 2014 Peter T. Coleman

Morton Deutsch with Peter Coleman and Exemplary Scholar-Practitioner in the Field of Social Justice, Gretchen Buchenholz at the 2013 Morton Deutsch Awards at Teachers College, Columbia University. *Source* From the author's personal photo collection

Morton Deutsch giving the Presidential Address at the Conference of the International Society of Political Psychology in 1982. *Source* From the author's personal photo collection

Acknowledgments

I wish to acknowledge the important and persistent helpful efforts of Elizabeth (Liz Hernandez) who enabled the publication of these two volumes. I also wish to thank Hans Günter Brauch the editor of this volume for his persistent, fruitful work arranging for the publication of these volumes. I also wish to acknowledge the help of Ariel Bernstein, Tiffany Brown, Ljubica Chatman, Christine T. Chung, Ginevra Drinka, Jen Hull, Leslie Migliaccio, Kristen Rucki, and Ling Shen, for their editing and proof-reading of the text. I also wish to acknowledge the work of many other psychologists who were colleagues in contributing to the psychological study of issues related to war and peace. They include, among many others: Jerome Frank, Charles Osgood, David Adams, Herbert Kelman, Dan Christie, David W. Johnson, Daniel Druckman, Dean G. Pruitt, Ronald J. Fisher, and especially Peter T. Coleman.

New York, December 2014 Morton Deutsch

Morton Deutsch with Dr. and Mrs. Issacharoff in a meeting of his book club, 2014. *Source* From the author's personal photo collection

Contents

Picture taken in summer of 2000 at the "Theory Conference" held at David Johnson's horse farm in Minnesota—*Frontrow* Roger Johnson, Dean Tjosvold, Peter Coleman—*Backrow* David Johnson, Norman Miller, Morton Deutsch, Frank Murray, Laurie Stevahn. *Source* The author's personal photo collection

Part I
On Morton Deutsch

Conference in Cancun, Mexico, 1979. *Front row* Alvin Zander, Dorwin Cartwright, Kurt Bach, Stanley Schachter, Leon Festinger, Elliot Aronson, Morton Deutsch—*Back row* Jerome E. Singer, Richard E. Nisbett. *Source* The author's personal photo collection. *Source* From the author's personal photo

Morton Deutsch and his wife Lydia about 1975. *Source* The author's personal photo collection

Chapter 1
Autobiography of Morton Deutsch: A Personal Perspective on the Development of Social Psychology in the Twentieth Century

Reflecting on his career as a social psychologist, Morton Deutsch guides us through a remarkable number of significant events that have shaped the field. He begins with his experience under the leadership of Kurt Lewin and the impact of the intellectual atmosphere that prevailed at the Research Center for Group Dynamics, which shaped not only his dissertation but also his entire value orientation as a social psychologist. He tells of his later work within the more applied atmosphere of the National Training Laboratory led by Ron Lippitt, describing his own particular research and many of the indelible contributions he has made to the field. Deutsch observes that his career as a social psychologist has centered on two continuing themes: cooperation, competition, and conflict on the one hand and distributive justice on the other. He concludes his reflections with the hope that future social psychologists will achieve a successful integration of three of the intellectual heroes of his youth: Freud, Marx, and Lewin.

My life almost spans the existence of modern social psychology.[1] My commentary on social psychology will be from the personal perspective of a reflection on my career as a social psychologist and the factors, social and personal, which influenced its development. However, I shall precede my autobiographical reflection with a brief commentary on the development of social psychology prior to my exposure to it.

Although modern social psychology was born in the first decades of the twentieth century, its ancestry in social philosophy can be traced back to ancient times. (For an excellent review of the precursors of modern social psychology, see Allport 1954a). It is a child of psychology and sociology, having been conceived in the ambivalent mood of optimism and despair that has characterized the scientific age. The rapidly expanding knowledge, the increasing confidence in scientific methods, the ever quickening technological change with its resulting opportunities and social problems, the development of new social organizations and of social planning, the social turmoil, the repeated disruption of communities and social traditions—all these helped to

---

[1] This text was first published as: "A personal perspective on the development of social psychology in the twentieth century. In Rodriguez, A. and Levine, R.V. (Eds.), *Reflections on 100 Years of Experimental Social Psychology* (New York, NY: Basic Books, 1999): 1–34. Permission to republish this text was granted by Ms. Isabelle Bleecker, Director, International Rights, Perseus Books Group, Boston, MA 02210 on 10 November 2014.

P.T. Coleman and M. Deutsch, *Morton Deutsch: A Pioneer in Developing Peace Psychology*, SpringerBriefs on Pioneers in Science and Practice 30,
DOI 10.1007/978-3-319-15440-4_1

create both the need for social psychology and the awareness of the possibility that scientific methods might be applied to the understanding of social behavior.

Charles Darwin's theory of evolution dominated the intellectual atmosphere of the time, and it became a model for early theorists in social psychology, who also set as their goal the achievement of a broad, encompassing theory of social behavior. The programmatic statements of theorists such as Charles Colley (1902), Tarde (1903), McDougall (1908), and Ross (1908) were grandly ambitious in scope but meager in detail. Many of the initial explanations of social behavior were made in terms of such processes as sympathy, imitation, and suggestion, which, in turn, were thought to be instinctually determined. The "herd instinct," the "instinct of submission," the "parental instinct," and a host of other instincts were invoked as innate, evolutionary derived causes of behavior.

The instinctual doctrines, however, did not last long. By the middle of the 1920s, they were in retreat. The prestige of the empirical methods in the physical sciences, the point of view of social determinism advanced by Karl Marx and various sociological theorists, and the findings of cultural anthropologists all contributed to their downfall. The two emphases in the rebellion against the instinctivist position, the rejection of the notion of instinctually caused behavior and the methodological stress on empirical procedures, still color contemporary social psychology. Empiricism is an inheritance from psychology; environmentalism is a legacy of sociology.

Opposition to the doctrine of instincts and, along with it, the minimization of genetic as compared to environmental influences upon social behavior led to many studies that illustrated the effects of social factors on individual psychological processes. [Bartlett's (1932) "Social Factors in Recall," Sherif's (1936) "Group Influences Upon the Formation of Norms and Attitudes," and Piaget's (1948) "Social Factors in Moral Judgment" are classic studies of this genre.] In consonance with the rapid social changes so characteristic of the modern period, investigations by social psychologists challenged long-held views about the fixity of human nature and about the innate superiority or inferiority of any social class, national group, or race. Social psychologists were not initially unsympathetic to Watson's (1930) extravagant assertion that "there is no such thing as inheritance of capacity, talent, temperament, mental constitution, and characteristics." More recently, there has been recognition that any full explanation of the development of human behavior must take into account the genetically determined biological equipment with which individuals confront their environment; even more lately, the emergence of "evolutionary social psychology" reflects this emphasis. Yet almost all social psychologists still reject the view of innate superiority-inferiority and the notion that social behavior is 'fixed' by instinct.

The rejection of abstract theorizing about social behavior in favor of empirical investigation provided the stimulus for the development of a variety of methods for studying social behavior; systematic interviews to obtain information about the motivations underlying behavior; controlled observational procedures to describe and classify behavior in social situations: methods of content analysis to analyze speeches, documents, and newspapers: sociometric techniques to study the social bonds and patterns of social interaction within a community; projective instruments of the study of personality patterns; and so forth. These methods have been extensively applied in public opinion polling, consumer research, studies of morale, investigations of prejudice and discrimination, personnel selection, and the like.

This revolt against armchair theorizing led many social psychologists not only to leave their armchairs but also to stop theorizing. Or perhaps it is more accurate to say that social psychologists who began to engage in empirical research in the 1920s and early 1930s did little to connect their research with theoretical ideas. During this same period, the psychoanalysts and also the early theorist abandoned their armchair mainly for the lecture podium.

Toward the end of the 1930s, under the enthusiastic but gentle leadership of Kurt Lewin, modern experimental social psychology began to flourish. Lewin and his students demonstrated that it is possible to create and study groups in the experimental laboratory that have important features in common with real-life groups. In doing so, they stimulated an interest in social psychological experimentation and attracted many experimentalists to work in this area.

Box 1.1: Morton Deutsch: Curriculum Vitae

Background

Born February 4, 1920
B.S., City College of New York, 1939;
M.A., University of Pennsylvania, 1940;
Ph.D., Massachusetts Institute of Technology, 1948;
Certificate in Psychoanalysis, Postgraduate Center for Mental Health, 1958.

Positions

Assistant Professor, New York University, 1948–1952; Associate Professor, New York University, 1952–1956; Member of Technical Staff (in charge of Interpersonal Process), Bell Telephone Laboratories, 1956–1963; Adjunct Professor, New York University, 1961–1963; Staff, Postgraduate Center for Psychotherapy, 1954–1963; Professor, Teachers College, Columbia University, 1963–1981; Edward Lee Thorndike Professor of Psychology and Education, Teachers College, Columbia University, 1981–1990; Director, International Center for Cooperation and Conflict Resolution, 1986–1998; Professor Emeritus, 1990–Present.

Professional Memberships

Fellow, American Psychological Association; Fellow, American Association for the Advancement of Science; New York State Psychological Association; Society for the Psychological Study of Social Issues; International Society of Political Psychology; Society of Experimental Psychology; European Association of Experimental Social Psychology; International Association of Conflict Management.

Areas of Specialization

Conflict resolution, distributive justice, social perception, intergroup relations, developmental social psychology, small group processes, social psychology of mental health.

Offices and Awards
President, Society for the Psychological Study of Social Issues (1961–1962); President, Division of Personality and Social Psychology of the American Psychological Association (1964–1965); President, New York State Psychological Association (1965–1966); President, Eastern Psychological Association (1968–1969); President, International Society of Political Psychology (1981–1982); first President, Division of Peace Psychology, American Psychological Association (1990–1991); consulting editorships: Journal of Experimental Social Psychology, Journal of Applied Social Psychology, Journal of Conflict Resolution, Journal of Personality and Social Psychology, Contemporary Psychology, Contemporary Psychoanalysis, Journal of Applied Behavioral Science, Journal of Peace Research, International Journal of Conflict Management, Peace and Conflict: Journal of Peace Psychology; the social psychological prize of the American Association for the Advancement of Science, 1961; the Samuel Flowerman Memorial Award of the New York Society for Clinical Psychologists, 1963; the Hovland Memorial Award Lectures, Yale University, 1967; the Kurt Lewin Award, the Society for the Psychological Study of Social Issues (SPSSI), 1968; Research Scientist Fellowship, National Institute of Mental Health, 1970–1971; the Gordon Allport Prize, SPSSI, 1973; Visiting Scholar, Russell Sage Foundation, 1976–1977; the Kurt Lewin Award of the New York State Psychological Association, 1980; the Cattell Fund Sabbatical Fellowship, 1983–1984; the Nevit Sanford Award of the International Society of Political Psychology, 1984; the Distinguished Scientist Award of the Society of Experimental Social Psychology, 1985; the Distinguished Scientific Contribution Award of the American Psychological Association, 1987; elected a William James Fellow of the American Psychological Society, 1988; Honorary Doctorate of Humane Letters by the City University of New York, 1989; the Helsinki Medal for Distinguished Contributions to Psychology by the University of Helsinki, 1990; the Teachers College Medal for Distinguished Contributions to Education, 1992; Distinguished Visiting Fellowship, La Trobe University, 1993; Life-time Achievement Awards: Psychologists for Social Responsibility, 1991; the Division of Cooperative Learning of the American Education Research Association, 1993; International Association of Conflict Management, 1993; Society for the Psychological Study of Social Issues, 1995; Society for the study of Peace, Conflict, and Violence, 1995; Levinson Award of the Division of Consulting Psychology of the American Psychological Association, 1998; the Book Award (for *the Handbook of Conflict Resolution: Theory and Practice*), The CPR Institute for Dispute Resolution, 2000; Annual Awards named the Morton Deutsch Award: The Society for the Study of Peace, Conflict, and Violence Award for distinguished work in the field of Conflict Resolution, 2003; The International Society for Justice Award for distinguished work in the field of social justice, 2004; Teachers College, Columbia University Award for a

distinguished graduate student paper related to social justice, 2005; Teachers College, Columbia University Award for distinguished scholarly/activist contributions to social justice.

1.1 Autobiography: Presocial Psychology

I was born, prematurely, in 1920 into a Jewish middle-class family in New York City, the last of four sons. I was always eager to catch up with my older brothers, feeling like an underdog, so I skipped through elementary and high school and entered the City College of New York (CCNY) in 1935 at the age of fifteen: two and a half years younger than most students.

I started off as a pre-med major with the idea of becoming a psychiatrist, having been intrigued by the writings of Sigmund Freud, some of which I read before college. I was drawn to psychoanalysis undoubtedly because it appeared to be so relevant to the personal issues with which I was struggling, and also because it was so radical and rebellious (it seemed to be so in the early and mid-1930s). During my adolescence, I was also politically radical and somewhat rebellious toward authority, helping to organize a student strike against the terrible food in the high school lunchroom and, later, a strike against the summer resort owners who were exploiting the college student waiters, of whom I was one.

The 1930s were a turbulent period, internationally as well as domestically. The economic depression, labor unrest, the rise of Nazism and other forms of totalitarianism, the Spanish civil war, the ideas of Marx, Freud, and Albert Einstein, as well as the impending Second World War were shaping the intellectual atmosphere that affected psychology. Several members of the psychology facility at CCNY were active in creating the Psychologist League, the precursor to the Society for the Psychological Study of Social Issues. Thus when I became disenchanted with the idea of being a pre-med student after dissecting a pig in a biology lab, I was happy to switch to a psychology major: It was a simpatico faculty. Psychology was a part of the Department of Philosophy at CCNY when I started my major in it. Morris Raphael Cohen, the distinguished philosopher of science, was the leading intellectual figure at CCNY, and his influence permeated the atmosphere.

At CCNY Max Hertzman introduced me to the ideas of Kurt Lewin and other Gestalt theorists. And under Walter Scott Neff's direction, I conducted my first laboratory experiment, a variation on Sherif's study of social norms, employing the autokinetic effect. As I now recall, in it I introduced a stooge who constantly judged the stationary speck of light in a dark room as having moved a substantial distance in one direction. (Most subjects see the light as moving a small difference in one direction.) The stooge has a considerable impact on the judgments made by the naïve majority of subjects. The findings of this pilot study anticipated later research by Serge Moscovici on minority influence.

My first exposure to Lewin's writings was in two undergraduate courses taken simultaneously: social psychology and personality and motivation. In the social psychology course, one of our textbooks was Brown's *Psychology and the Social Order* (1936). This was an ambitious, challenging, and curious text that tried to apply to the major social issues of the 1930s Lewinian and Marxian ideas, with a sprinkling of the Riemanian geometry employed by Einstein in his theory of relativity. To a naïve 17-year-old undergraduate student like me, it was a very impressive and inspiring book showing how social science could shed light on the urgent problems of our time.

Kurt Lewin (1890–1947). *Source* "Biographies and lifes" (Spanish), at: <http://www.biografiasyvidas.com/biografia/l/lewin.htm> and at: "Great Thoughts Treasury"; at: <http://blogs.scientificamerican.com/moral-universe/files/2013/08/lewin.jpg>

In the personality and motivation course, I read Lewin's *Dynamic Theory of Personality* (1935) and *Principle of Topological Psychology* (1936). I also read his *Conceptual Representation and Measurement of Psychological Forces* (1938) as an undergraduate, but I cannot recall when. I and others experienced great intellectual excitement on reading these books more than 50 years ago. *A Dynamic Theory of* Personality consisted of a collection of independent articles, previously published in the early 1930s, whereas the other books made a brilliant but flawed attempt to articulate the foundations of a scientific psychology with the aid of topology. They were mind openers. These books are permeated by a view of the nature of psychological science different from what was then traditional. The new view was characterized by Lewin as the "Galilean mode of thought," which contrasted with the classical "Aristotelian mode." In my writings on field theory (Deutsch 1968), I have characterized in some detail Lewin's approach to psychological theorizing, his metatheory.

Although I was impressed by Lewin's writings, my career aspirations in psychology were still focused on becoming a psychoanalytic psychologist as I decided to do graduate work in psychology. My undergraduate experiences, in as well as outside the classroom, led me to believe that an integration of psychoanalysis, Marxism, and scientific method, as exemplified by Lewin's work, could be achieved. In the 1930s, such influential figures as Wilhelm Reich, Erich Fromm, Max Horkheimer, Theodor Adorno and Else Frenkel-Brunswik, as well as many others, were trying to develop an integration of psychoanalysis and Marxism. Also at this time, some psychoanalytic theorists such as David Rappaport were intrigued by the idea that research conducted by Lewin and his students on tension systems could be viewed as a form of experimental psychoanalysis.

I am not sure why I was advised to go to the University of Pennsylvania to take my master's degree. Possibly it was because it had a well-established psychological clinic and two faculty members, Frances Irwin and Malcolm Preston, who were sympathetic to Lewin's ideas. I had some interesting clinical experiences there working with children, largely without supervision, but the coursework seemed dull and antiquated in comparison with my undergraduate courses at CCNY. I earned the reputation of being a radical by challenging what I considered to be racist statements about Negro intelligence in a course on psychological measurement given by Morris Viteles.

After obtaining my M.A. degree in 1940, I started a rotating clinical internship at three New York state institutions: one was for the feebleminded (Letchworth Village), another for delinquent boys (Warwick), and a third for psychotic children as well as adults (Rockland State Hospital). During my internship, I became skilled in diagnostic testing and clinical interventions with a considerable variety of inmates, more widely read in psychoanalysis, and more aware of how some capable inmates were unjustly retained in the institution because of the valuable services they performed for it or its staff.

I also had the good fortune to meet Clark Hull (the famous learning theorist) while he was visiting a former doctoral student of his, a staff psychologist at Letchworth Village. He was a remarkably generous and tolerant person. We had several long discussions, one related to his recently published book developing a hypothetico-deductive system for rote learning. I had read the book and was somewhat critical of it from two perspectives: the perspective of Gestalt psychology and of Morris Cohen and Ernst Nagel's book on scientific method, both of which I had been thoroughly indoctrinated in while I was an undergraduate at CCNY. Hull seemed genuinely interested in what I had to say even though I was an overly brash 20-year-old pipsqueak. We had another interesting discussion in which he gave me advice on how to seduce a woman. He told me that, on a date, I should carry a handkerchief permeated with perspiration. He explained that sweat and sexual feelings were associated together because of their joint occurrence during sexual intercourse and that sweat would arouse sexual feelings. In retrospect, I realize that he must have been joking since his suggestion never worked for me.

When Pearl Harbor occurred in December 1941, I was still in my psychology internship. Shortly thereafter, I joined the air force. My first assignment was to a psychological research unit at Maxwell Field in Alabama, which did psychological testing of aviation cadets to classify them for training as pilots, navigators, or bombardiers. I soon became bored with testing and wanted to participate directly in action against the Nazis. I became a cadet and was trained as a navigator. To get to our combat base in England, our crew flew to and stopped at bases in such exotic sport as Trinidad: Fortaleza and Belem in Brazil; Dakar and Marrakech in Africa; and Scotland. What an eye-opening cross-cultural experience; I had never been outside the Northeastern part of the United States before joining the air force.

I flew in thirty bombing missions against the Germans. During combat, I saw many of our planes as well as German planes shot down, and I also saw the massive damage inflicted by our bombs and those of the Royal Air Force on occupied Europe and Germany. Moreover, being stationed in England, I saw the great destruction wreaked by the German air raids and felt the common apprehensions while sitting in air-raid shelters during German bombings. Although I had no doubt of the justness of the war against the Nazis, I was appalled by its destructiveness.

After my combat tour of duty was completed, I returned to the United States and was assigned as a clinical psychologist to an Air Force Convalescent Hospital and served as such until shortly after V-E Day. I was demobilized early as the result of being one of the few non-patients at the hospital who had been in combat and had amassed a substantial number of demobilization points.

After my demobilization, I contacted some psychology faculty members I knew at CCNY to ask for advice with regard to resuming graduate work in psychology; I discussed with them my somewhat confused interests in getting clinical training, in studying with Lewin because of his work on democratic and autocratic leadership, and in doing psychological research. As a result of these conversations, I decided to apply for admission to the doctoral programs at the University of Chicago (where Carl Rogers and L.L. Thurstone were the leading lights), at Yale University (where Donald Marquis was chairman and where Clark Hull was the major attraction), and at MIT (where Kurt Lewin had established a new graduate program and the Research Center for Group Dynamics). As one of the first of the returning soldiers, I had no trouble in getting interviews or admission at all three schools. I was most impressed by Kurt Lewin and his vision of his newly established research center and so I decided to take my Ph.D. at MIT.

1.2 My Autobiography as a Social Psychologist

I date the start of my career as a social psychologist to my first meeting with Lewin, in which I was enthralled by him and committed myself to studying at his center. He had arranged for me to meet him for breakfast at a midtown hotel in New York in August 1945. Even though it was very hot, I dressed formally—with jacket and tie—to meet with this distinguished professor. Our meeting time was 8:30 A.M., but he did not appear until about 9:00 A.M. He came bustling in, cheerfully looking around for me, his face bright pink from a recent sunburn. He was not wearing a jacket or a tie, and his manner was quite informal. I recognized him from a picture that I had seen and introduced myself, and we set off for the hotel's dining room. However, they would not admit us because he had no jacket or tie (how things have changed). We then went to a nearby coffee shop. I do not remember much about the conversation other than that I described my education, experience, and interests, and he described his plans for the new center. I was experiencing a trancelike sensation of intellectual illumination with new insights constantly bubbling forth from this brilliant, enthusiastic, effervescent, youthful, middle-aged man. He spoke a colloquial American, often with malapropisms, and he was both endearing and charming. I left the interview with no doubt that I wanted to study with Lewin. I also left in a dazed sense of enlightenment, but I could not specifically identify what I was enlightened about when I later tried to pin it down for myself.

I had a similar experience a month later when I went to MIT to study and work with Lewin. He discussed with me some work he was then doing with the Commission on Community Interrelations of the American Jewish Congress (a commission he helped to establish) to reduce anti-Semitism and other forms of prejudice. His discussion of the issues was intensely illuminating when I was with him, but I could not define it afterwards when I was alone. At the end of our meeting, he asked me to prepare a review of the essence of the literature on prejudice, and he indicated that it should be brief and that he needed it in 3 days. I felt good. I was being treated as a serious professional and was given a responsible and challenging task. Lewin's treatment of me was, I believe, typical of his relations with his colleagues and students. He would discuss a topic with great enthusiasm and insight, he would ignite one's interest, and he would encourage one to get involved in a task that was intellectually challenging, giving complete freedom for one to work on it as one saw fit.

Shortly after arriving at MIT; I noticed a very attractive young woman named Lydia Shapiro who would occasionally pop into the center. She was working under Lewin's direction as an interviewee for a study on self-hatred among Jews. We started to get to know one another over cherry Cokes and jelly donuts. Being supported on the GI bill, I was a cheapskate, and she did like jelly donuts. I don't recall the specifics, but somehow I was assigned to supervise her work. After learning that she spent much of her supposed work time sunning herself on the banks of the Charles River, I fired her. About a year and a half later, on June 1st, 1947, we got married. Stan Schachter and Al Pepitone, with whom I was sharing an apartment, were my best men at the wedding. In moments of marital tension, I have accused Lydia of marrying me to get even, but she asserts it was pure masochism on her part. In our 50 years of marriage, I have had splendid opportunities to study conflict as a participant observer.

Immediately after our honeymoon in Quebec, we went to Bethel in Maine for the first National Training Laboratory (NTL). I served on its research staff with other students from the RGGD at MIT and from the Harvard Department of Social Relations. Lydia and another woman were the rumrunners for the workshop; Bethel was a dry town, and they had to drive 20 miles to buy the liquor to keep the workshop staff and participants well lubricated.

The first NTL was a natural follow-up of the Connecticut Workshop on Inter-group Relations held during the summer of 1946. As I now recall it, the training staff consisted of Ron Lippitt, Ken Benne, and Lee Bradford, and the research staff consisted of Murray Horowitz, Mef Seeman, and myself. One evening, following a lengthy workshop day, Lewin, the workshop participants, the trainers, and the researchers were all sitting around a conference table when one of the participants turned to the researchers and asked us what we were doing. We said that we were keeping track of the patterns of interaction among the group. He then asked us to describe what we had noted; Lewin suggested that it would be an interesting thing to do. We summarized our impressions, and this lead to a lively, insightful learning experience. This was the embryo of the T-group and sensitivity training that was given birth at the first NTL in 1947.

I would now say that the researchers at the first NTL did not fully appreciate the importance of the new procedures and new movement being developed. The evangelical tone of some of the trainers appalled many of us, with the result that there was considerable unhappiness among the researchers that summer of 1947. Today many of us recognize that NTL as the birthplace of much of applied social psychology, especially in the area of organizational psychology.

1.3 The Research Center for Group Dynamics

Lewin assembled a remarkable group of faculty and students to compose the Research Center for Group Dynamics at MIT. For the faculty, he initially recruited Dorwin Cartwright, Leon Festinger, Ronald Lippitt, and Marian Radke (now Radke-Yarrow). Jack French and Alvin Zander were to join later. The small group of twelve students included Kurt Back, Alex Bavelas, David Emery, Gordon Hearn, Murray Horowitz, David Jenkins, Albert Pepitone, Stanley Schachter, Richard Snyder, John Thibaut, Ben Willerman, and myself. These initial faculty and students were extraordinarily productive, and they played a pivotal role in developing modern social psychology in its applied as well as its basic aspects. As I write these last two sentences, it strikes me that all of the students and the key faculty members were male. This was quite a change for Lewin; in Berlin, most of his students were female (e.g. Bluma Zeigarnik, Tamara Dembo, Eugenia Hanfmann, Maria Ovsiankina, Anitra Karsten). It is interesting to speculate how modern social psychology's development might have differed if the student group included a substantial number of women.

Lewin died suddenly on February 11, 1947 of a heart attack. The RCGD had been functioning for considerably less than 2 years when he died. Yet in this brief period of time he had established an institution that would strongly influence the development of modern social psychology. Let me offer some thoughts about why the Research Center for Group Dynamics was so remarkably productive.

1.3.1 Reasons for the Center's Effectiveness

First, Lewin was an unusually effective scientific "tribal leader" (to borrow a phrase from Donald Campbell). As I have indicated in describing my personal contacts with him, he was enthusiastic, inspiring, and persuasive. He led those working with him to feel they were involved in an important, promising enterprise that could have valuable consequences for both social science and society. He treated his faculty and students as colleagues: giving them autonomy and responsibility and a sense of being actively involved, individually and collectively, in creating the new field of group dynamics. He also encouraged open and vigorous conflict about ideas and methods among his faculty and students in the never ceasing attempt to get to a deeper understanding of the issues involved.

This was most evident in the loosely organized research seminars, named the Quasselstrippe (or winding string), he led for the faculty and students. In the Quasselstrippe, a faculty member or student would typically present some research or some theoretical issue that he or she was involved in, and a lively controversy would erupt. Sometimes the controversy was related to the presentation, but frequently the discussion wandered off into other issues. Not infrequently, the most heated exchanges took place between Leon Festinger and Ronald Lippitt, who had rather different views of the nature of science and research. During these vigorous disputes, Lewin would be smiling benignly as he watched his intellectual offspring squabble. Almost invariably at the end of these wandering, disputatious research seminars he would emerge from his role as an observer, and in an active way he would offer a deeper, integrating perspective that would provide a basis for synthesizing the conflicting viewpoints.

It was not only Lewin's leadership style but also his ideas that contributed to the productivity of the RCGD. Very much influenced by Ernst Cassirer, the German philosopher of science, he thought, "the taboo against believing in the existence of a social entity is probably most effectively broken by handling this entity experimentally" (Lewin 1951:193). The concept of "group" as well as other concepts relating to social psychological phenomena had little scientific status among psychologist in the 1930s and 1940s when Lewin was first turning his attention to social psychology. He believed the 'reality' of these concepts would be established only by "doing something with them." So at the center there was strong pressure to do something with the concepts related to groups and not merely to talk about these ideas.

And, of course, the faculty and students did many experiments to demonstrate that one could, in a sense, capture for science such phenomena as "styles of group leadership," "social influence," "cooperation and competition," "group cohesiveness," "pressure for uniformity," "trust and suspicion," "social comparison," and so on. The pressure to do something with the concepts was directed not only toward experimentation but also toward application, namely, to show that these concepts could be employed to change exiting social reality—to improve group functioning, to reduce prejudice, or to train more effective leaders.

Lewin's metatheory, his conceptual language, as well as his specific theoretical ideas were also important influences on the members of the center while they were at MIT. More than 30 years later, in the spring of 1978, there was a reunion at Columbia University of almost all of the surviving RCGD members. The participants included Kurt Bach, Dorwin Cartwright, Leon Festinger, Jack French, Gordon Hearn, Harold Kelley, Ronald Lippitt (via tape), Albert Pepitone, Stanley Schachter, and myself. At that reunion, the participants were asked to indicate Lewin's effect on their work. From the discussion, it was evident that all of us had been very much influenced by Lewin's way of thinking about science and by his general orientation to psychology. Elsewhere I have described the key elements of Lewin's metatheory—in other words, his field-theoretical approach to psychology. This is what had most impact on the participants. Few were still involved in Lewin's conceptual language or terminology, with topological and vectorial psychology. Some had been stimulated to do work that related to Lewin's specific theoretical ideas, particularly those relevant to tension system, level of aspiration theory, social interdependence, group leadership, group decision making, changing individual attitudes, and quasi-stationary equilibria. Several were stimulated by Lewin to be concerned with articulating the connection between social psychology theory and change in social practice.

Nevertheless, the common thread that linked our group of past RCGD members together was a Lewinian way of thinking. It emphasized the importance of theory; the value of experimentation for clarifying and testing ideas; the interrelatedness between the person and the environment; the interdependence of cognitive structures and motivation; the importance of understanding the individual in his or her social (group, cultural) context; the usefulness of theory for social practice; and the value of trying to change reality for the development of theory. These emphases are not unique to the Lewinian way of thinking; they characterize good social science and good social practice. Lewin was the one who introduced them to social psychology.

The RCGD fostered a sense of pioneering elitism among its members. We felt we were working on the frontiers of social psychological knowledge, creating new research methods, and capturing new phenomena for science. This fostered a narcissistic arrogance in many of us that permitted us to venture on untrodden paths and to feel rather superior to the work being done by our friends and neighbors in Harvard's Social Relations Department as well as elsewhere.

In addition, of course, the center had a critical mass of active researchers among its faculty and students, so that the publications of this group dominated the early work in experimental and applied social psychology. Marrow (1969), in his biography of Kurt Lewin (*The Practical Theorist*), listed over 100 publications and dissertations connected with the RCGD during the period of 1945–1950. In a sense, apart from whatever merits we had, we were so influential because we were lucky enough to be active early in the development of modern social psychology when there were comparatively few others who were doing research and publishing in this field.

Lewin recruited a very able and congenial group of mature students who, for the most part, had done previous graduate work in psychology and had served in the armed forces in World War II. They were prepared to take responsibility and to work with the faculty as colleagues. The relatively young faculty members were unusually accessible and open to collaborative working relations with the students. As students, we were quickly involved in the design and execution of experiments and research on training workshops; some of us were also rapidly thrust into the role of conducting training workshops on group processes and group leadership. The students comprised a small, cohesive group that provided much mutual support even as we had intense intellectual discussions about the new ideas and techniques that were being developed.

Lewin also recruited a remarkably gifted younger faculty. I assume that he purposefully created a faculty that had some tension as well as some unifying elements within it, a faculty within which there would be productive tension in theory, research, and application. As suggested earlier, Festinger and Lippitt had fundamental disagreements, and while he lived, Lewin served as an integrating force, intellectually as well as administratively. After his death, Cartwright maintained administrative integration, but there was little intellectual common ground between the disparate perspectives of Festinger and Lippitt. For many students, Festinger became a symbol of the tough-minded, theory-oriented, pure experimental scientist, whereas Lippitt became a symbol of the fuzzy-minded, do-gooder, practitioner of applied social psychology. These were unfortunate caricatures of both Festinger and Lippitt. Such distortions were, I believe, one of the contributing causes to the estrangement between basic and applied social psychology in the United States during the 1950s and early 1960s. I doubt that these caricatures would have developed if Lewin had lived longer. As my earlier quotation from him indicated, he saw an intimate, two-directional link between the development of theory and practice.

My career in social psychology has been greatly affected by Kurt Lewin and my experiences at the Research Center for Group Dynamics.[2] First, I probably would not have been a social psychologist were it not for the inspiring interview with him in the summer of 1945. Second, the intellectual atmosphere created by Lewin at the RCGD strongly shaped my dissertation and my value orientation as a social psychologist.

Lewin was not only an original, tough-minded theorist and researcher with a profound interest in the philosophy and methodology of science, but he was also a tenderhearted psychologist who was deeply involved with developing psychological

---

[2] Lewin was widely admired by other psychologists. In the summer of 1947, after his death, there was a meeting of the Topological Circle at Smith College. At this meeting there were such eminent psychologists as Fritz Heider (the host), Edward Chace Tolman, and David Rappaport, as well as many of the faculty and students of the RCGD. At that meeting Heider presented the ideas that are the core of his subsequently published book. Heider was a shy and somewhat inarticulate public speaker, but the profundity of his ideas gripped us all. The meeting also provided us the opportunity to have lively informal discussion with Tolman and Rappaport (who offered me a job at Austen Riggs).

knowledge that would be relevant to important human concerns. Lewin was both tough-minded and tenderhearted; he provided a scientific role model that I have tried to emulate. Like Lewin, I have wanted my theory and research to be relevant to important social issues, but I also wanted my work to be scientifically rigorous and tough-minded. As a student, I was drawn to both the tough-mindedness of Festinger's work and to the direct social relevance of Lippitt's approach and did not feel the need to identify with one, derogate the other.

1.4 My Dissertation Study

My dissertation started off with an interest in issues of war and peace (atomic bombs had been dropped on Hiroshima and Nagasaki shortly before I resumed my graduate studies) and with an image of the possible ways that the nations composing the newly formed United Nations Security Council would interact. The atmosphere at the center, still persisting after Lewin's premature death, led me to turn this social concern about the risk of nuclear war into a theoretically oriented, experimental investigation of the effects of cooperative and competitive processes. The specific problem that I was first interested in took on a more generalized form. It had been transformed into an attempt to understand the fundamental features of cooperative and competitive relations and the consequences of these different types of interdependencies in a way that would be generally applicable to the relations among individuals, groups, or nations. The problem had become a theoretical one, with the broad scientific goal of attempting to interrelate and give insight into a variety of phenomena through several fundamental concepts and basic propositions. The intellectual atmosphere at the center pushed its students to theory building. Lewin's favorite slogan was, "there is nothing so practical as a good theory."

As I reflect back on the intellectual roots of my dissertation, I see it was influenced not only by Lewin's theoretical interest in social interdependence but also by the Marxist concern with two different systems of distributive justice: a cooperative, egalitarian and a competitive, meritocratic one. In addition, the writings of George Herbert Mead affected my way of thinking about cooperation and its importance to civilized life.

This study,[3] in addition to being the takeoff point for much of my subsequent work, has helped to stimulate the development of a movement toward cooperative

---

[3] One sour note in connection with my dissertation: For it, I had developed an observation schedule and manual describing the "function of participation" for characterizing the behavior of group members. It included a description and detailing of various task, group, and individual functions. I also used this material in analyzing observational data in connection with the research done on the first NTL. Much to my surprise, shortly before my dissertation defense in the summer of 1948, an article by Kenneth Benne and Paul Sheats entitled "The Functional Role of Group Members" appeared in the *Journal of Social Issues*. This article was mainly a reprint of my manual with some elaborations; my authorship received no acknowledgement. When I brought this to the

learning in the schools under the leadership of David and Roger Johnson. Although cooperative learning has many ancestors and can be traced back for at least 2,000 years, my dissertation helped to initiate the development of a systematic theoretical and research base for cooperative learning. Hundreds of research studies have since been done on the relative impact of cooperative, competitive, and individualistic learning (see Johnson/Johnson 1989). These various studies are quite consistent with one another and with my initial theoretical work and research on the effects of cooperation-competition (Deutsch 1949a, b) in indicating favorable effects upon students. Through cooperative learning, students develop a considerably greater commitment, helpfulness, and caring for one another regardless of differences in ability level, ethnic background, gender, social class, and physical ability. They develop more skill in taking the perspective of others, emotionally as well as cognitively. They develop greater selfesteem and a greater sense of being valued by their classmates. They develop more positive attitudes toward learning, school, and their teachers. They usually learn more in the subjects they are studying by cooperative learning, and they also acquire more of the skills and attitudes that are conducive to effective collaboration with others.

1.5 The Research Center for Human Relations

After obtaining my Ph.D. from MIT in the summer of 1948, I joined the Research Center for Human Relations (then at the New School) headed by Stuart Cook. The war against Nazism had stimulated a considerable interest among psychologists in understanding prejudice and how to overcome it, and financial support for research in this area was available form Jewish organizations such as the American Jewish Congress as well as from federal agencies. Among the many groups receiving funding for work in this area were members of the Berkeley Public Opinion Study and the former Frankfurt Institute of Social Research, who produced *The Authoritarian Personality* (Adorno et al. 1950); Lewin's MIT Center which developed not only the first workshop for reducing prejudice and improving intergroup relations but also action research "to help social agencies that were developing programs aimed at reducing prejudice and discrimination"; and the Harvard group working with Allport (1954b) on creating an integrated overview of the nature of prejudice and ways of reducing it.

The Research Center for Human Relations was, in 1948, also mainly funded by agencies interested in reducing prejudice. As soon as I joined, I became involved in

(Footnote 3 continued)
attention of Benne and Sheats, they acknowledged that their article was based on my manual, but since it did not have my name on it, they thought it was some impersonal product of NTL. They apologized for their error, but when the article was widely reprinted in books, there was no attempt to undo their error. When I published my dissertation, I included a footnote indicating that some of my dissertation material had been published in "The Functional Role of Group Members.".

a study of interracial housing that I conducted with Mary Evan Collins. We started with an "experience survey" of knowledgeable public housing officials to identify the important factors affecting interracial relations in housing projects. On the basis of this survey, we decided that the residential pattern—whether the races were segregated or integrated with in the housing project—was a critical determinant. We then set out to identify housing projects that were otherwise similar but differed in terms of whether black and white residents lived in separate buildings or were integrated within each building. We were able to identify biracial segregated public housing developments in Newark, New Jersey, and racially integrated ones in New York City that were roughly similar. We then did an extensive interview and a small observational study in the projects, and by the use of various controls, we created a quasi-ex post facto experiment. Despite the obvious methodological limitations of such a study, it was clear that the two types of projects differed profoundly in terms of the kinds of contacts between the two races and the attitudes that they developed toward each other.

This study (Deutsch/Collins 1951) had important social consequences. As the executive director of the Newark Public Housing Authority stated in a postscript to our book, *Interracial Housing*, "The partial segregation which has characterized public housing in Newark will no longer obtain. In large measure, this change in fundamental policy reflects the impact of the study reported in this book. The study has served as a catalyst to the re-examination of our basic interracial policies in housing and as a stimulus to this change." It also led me to become active on a Society for the Psychological Study of Social Issues (SPSSI) committee concerned with intergroup relations. Over the next several years, this committee gave talks before policy-oriented groups as well as helped lawyers who were challenging racial segregation in various suits brought before federal courts. The committee also contributed material to the legal brief that was cited in the 1954 Supreme Court decision *Brown v. the Board of Education*, which outlawed racial segregation in schools and other publicly supported facilities.

In 1949, the Research Center for Human Relations moved to New York University (NYU), and I became a member of its graduate faculty in psychology. Here, I worked collaboratively with Marie Jahoda and Stuart Cook on an SPSSI-sponsored textbook, Research Methods in Social Relations (Jahoda et al. 1951), one of the earliest—if not the earliest—of its kind. To help me overcome my Kafkaesque, Germanic style of writing, Mitzi pinned in my wall a slogan that stated, "You don't have to write complex sentences to be profound." It was a good reminder as well as a subtle way of deflating my pompous persona of theorist-basic researcher with which I had emerged from my graduate studies.

At NYU, I also worked collaboratively with Harold Gerard on a laboratory study of normative and informational influence on individual judgment (Deutsch/Gerard 1955) and a study of decision-making among high-level air force officers. In addition, with support from the Office of Naval Research, I was able to start a program of research on factors affecting the initiation of cooperation. Hal had introduced me to Howard Raffia, who in turn introduced me to the Prisoner's Dilemma (PD), which I soon turned into a useful research format for investigating trust and suspicion (Deutsch 1958, 1962a, 1973). I was probably the first psychologist to use the PD game in research. Unfortunately, the PD game (like the Asch situation and the Skinner box) became an easy format for conducting experimental studies, and as a result a torrent of studies followed—most of which had no theoretical significance.

I added to my busy schedule by undertaking training as a psychoanalyst at the Postgraduate Center for Mental Health, which had an eclectic orientation rather than being committed to one or another school of psychoanalysis. It involved not only my own analysis, (three times per week), but also 6–9 h of classes, 20 h of doing psychoanalytic psychotherapy, and 2–3 h of supervision per week. It was hectic, but I was young. It was an extremely valuable supplement to my work as an experimental social psychologist, which gives perspectives only on very narrow cross-sections of people's lives. Psychoanalysis provided a longitudinal, developmental view in addition to glimpses into the internal psychodynamics underlying a person's behavior in conflict situations. My psychoanalytic work stimulated my research interest in such topics as trust and suspicion and conflict. It has been a two-way street. My social psychological work on conflict, negotiation, and mediation has affected my therapeutic approach to the conflicts experienced by patients as well as my approach to marital therapy. I continued a small private practice until about 10 years ago, when I wanted to have more freedom to travel. The practice was personally rewarding. I helped a number of people, it enabled me to stay in touch with my own inner life, and it provided a welcome supplement to my academic salary.

Morton Deutsch with his first child (Anthony) in 1950. *Source* The author's personal photo collection

During my tenure at New York University, most of my salary was paid out of soft money, from research grants or other monies from outside sources. As McCarthyism developed increasing strength in the early 1950s, social science and social scientists became targets of attack, being labeled as "radical," "fellow travelers," "communist sympathizers," and the like. If your personal library contained books by Karl Marx, if you had participated in interracial groups challenging segregation, if a friend was or had been a member of the Communist Party, and so on, you were suspect and might be purged from your position. During the height of the McCarthy period, many funding agencies no longer were willing to support research dealing with prejudice or interracial relations, and there was much talk of reducing federal support for social science research. Thus I was happy to accept when Carl Hovland, in 1956, invited me to help establish a new basic research group in psychology at the Bell Telephone Laboratories. Bell Labs had an excellent reputation for its support for basic research, and this is what I wanted to do without the constant problem of raising money.

Much to my surprise, even during the worst part of McCarthyism I never had any problems, nor did my funding from the Office of Naval Research or the air force stop. Although never a communist, I had many of the characteristics of the "usual suspect." Possibly, I was not harassed because I had received a security clearance from the air force before doing research on decision-making in the early 1950s.

1.6 The Bell Laboratories

Bell Labs was, by academic standards, a luxurious place to work. I received a good salary and had no trouble getting research assistants, equipment, secretarial help, and travel money as well as much freedom to do what I wanted. I was able to hire Bob Krauss and Norah Rosenau, then graduate students at NYU, to work as my research assistants. I was also able to add Hal Gerard and Sy Rosenberg to our research staff. It was a productive group. At Bell Labs, Bob Krauss and I developed and conducted research with the Acme-Bolt Trucking game; we also started on our book, *Theories in Social Psychology* (Deutsch/Krauss 1965). I did various other studies including "The Interpretation of Praise and Criticism" (Deutsch 1961), "Dissonance or Defensiveness" (Deutsch et al. 1962), and "The Effects of Group Size and Task Structure upon Group Process and Performance (Deutsch/Rosenau 1963). This last was a fine study that was never written up for publication because of Norah Rosenau's premature death and my change of interests as I moved to Teachers College in 1963.

In addition, while at the Bell Labs, I was its unofficial peacenik, criticizing the strategic thinking among establishment intellectuals and coediting the book Preventing World War III (Wright et al. 1962). During this period I was quite active in SPSSI, articulating some of the social psychological assumptions underlying our national policy and even becoming its president.

Although Bell Labs was in many respects a fine place to work, it had its problems. Compared to a university, it was a stiff organization: It had a clear hierarchical structure; it had fairly set hours of work and vacation (from which I was a tolerated deviant); the lab had no small, offbeat, informal eating places that served wine or beer; there were few students and little ethnic and racial diversity.

In addition, there were specific problems related to our psychological research unit. Although it was located in the Bell Labs in Murray Hill, New Jersey, the Personnel Research Group at AT&T had been instrumental in getting the unit established and thought that we should be primarily working closely with them on problems with which they needed help. None of us who had come to Bell Labs at Carl Hovland's urging had this view, nor apparently did Carl. The administrative head of our unit was a former member of the AT&T Personnel Group. An uncomfortable power struggle developed about what we should be doing, which Bell Labs ultimately won. But because of the dispute and also because we were the oddballs of the Bell Labs (which was composed mainly of physical scientists and mathematicians), we were the constant object of high-level attention. We had visits from the president of AT&T, the president of Western Electric, the presidents of various Bell Telephone Companies, and so on, and at each visit our group would have to put on a show, lasting 1 or 2 days, in which we would demonstrate our research. During one of these visits, when a committee came in order to make a recommendation about the future of our group, we received word that Bob and I had just been awarded the American Association for the Advancement of Science (AAAS) sociopsychology prize for the research we had done at the Bell Labs with the Acme-Bold Trucking game (Deutsch/Krauss 1962). This apparently laid to rest the doubts about our group.

In addition to the people I recruited for my research group on interpersonal processes, Alex Bavelas, another key staff member selected by Hovland, recruited Herbert Jenkins, a Skinnerian who did his research on learning using pigeons. Herb must have had several people a day ask him, jokingly, "Going to replace the telephone with pigeons, eh?" After a year or so, Bavelas quit the labs, feeling that it was not a receptive environment for what he wanted to do. Jenkins then recruited Roger Shepard, who started his brilliant work on multidimensional scaling there.

While at the labs, I was consulted by its administration on problems such as how to improve the creativity of their researchers, how to apply social science knowledge to improve the functioning of the various telephone companies, and how to improve race relations. As I recall, I gave many potentially useful suggestions, none of which were implemented. I also suggested that they hire Henry Riecken to establish a social science development group to develop existing social science knowledge for use in the Bell system. Although Bell interviewed Riecken, they did not implement this idea either.

Hovland died in 1961, and about a year later I started to think about leaving the labs. I was getting tired of commuting from New York City to Murray Hill: I missed working with graduate students as well as the looser, less hierarchical atmosphere of a university; and I was bored by the special attention that our group was receiving. My memory of the specifics is unclear, but around this time I was approached by Teachers College to consider an appointment to replace Goodwin Watson, who was retiring, and to head its doctoral program in social psychology. Teachers College was attractive to me because Lydia and I were determined to continue living in New York, I would have freedom to create a new social psychology program, and I was interested in education. I received other feelers from nearby institutions (the Department of Management at Yale University and the Department of Psychiatry at the Albert Einstein College of Medicine) that would have provided higher salaries and more affluent settings, but they did not have the lure of shaping a social psychology program.

1.7 Teachers College

When I joined Teachers College in September 1963, I had a strong view of what I wanted the new social psychology program to be like. I wanted it to attract students and turn out graduates who would be tough-minded and tenderhearted, who would be as knowledgeable and expert in theory and research as the best of the 'pure,' experimental social psychologists and also socially concerned with developing and applying social psychological knowledge to the urgent and important social problems of our time. In other words, I wanted to develop a program that would overcome the split that had developed between the laboratory and applied social psychology during the 1950s and the early 1960s. As I have indicated earlier, the differences between the sharp-minded and sharp-tongued Festinger and the evangelical, unsystematic Lippitt were precursors of this split, which widened into a chasm in the decade after Lewin's death (see Deutsch 1975 for a more extensive discussion of this rift).

Although the split was understandable in terms of the insecurities of both sides in a young discipline, it was harmful and stupid from my perspective. It polluted the atmosphere of social psychology. When I left Bell Labs (a tough-minded institution) to join Teachers College (a tenderhearted one), I thought that my experimental colleagues would consider this to be a loss of status for me and that my new colleagues would be concerned that I would be overly critical and scientistic (rather than scientific) as well as out of touch with practical realities. However, by the time I came to Teachers College, I felt sufficiently secure in my own identity as a social psychologist not to be concerned by colleagues who would deprecate either tenderheartedness or tough-mindedness.

I was fortunate when I came to Teachers College in several respects. First, although Teachers College, like most schools of education, has relatively little money for research by its faculty or stipends for its graduate students, I was able to bring in outside funding to get the social psychology program off to a good start: The National Science Foundation (NSF) gave funds to build a well-equipped social psychology laboratory, the Office of Naval Research (ONR) supported my research, and the National Institute of Mental Health (NIMH) provided a training grant that would support most of our graduate students. Second, we were able to attract many excellent students who fit our criteria of being tough-minded and tenderhearted, including Harvey Hornstein, David Johnson, Jeffrey Rubin, Roy Lewicki, Barbara Bunker, Madeleine Heilman, Kenneth Kressel, Charles Judd Jr., Janice Steil, Michelle Fine, Ivan Lansberg, Louis Medvene, Susan Boardman, Sandra Horowitz, Susan Opotow, Even Weitzman, Martha Gephart, and Adrienne Asch. Third, our program was initially small enough for us to be a very cohesive group that mainly worked cooperatively on interrelated research projects under my direction. We could have frequent informal lunches together during which we discussed politics, diets, Jackie Ferguson (our fascinating secretary who mothered us all), and research and theory. Many good ideas emerged from these lunches. Finally, the change from Bell Labs to Teachers College accelerated a shift in focus and labeling of my research. At the Bell Labs, I and others came to view the Acme-Bolt Trucking game as a bargaining game, so I began to think of studies that employed it as bargaining or negotiation and more generally as conflict studies. This was a shift away from labeling them as studies of the condition affecting the initiation of cooperation.

With a change in labeling, I began to reframe the question underlying much of my research from "What are the conditions that give rise to cooperation rather than competition?" to "What are the conditions that give rise to constructive rather than destructive processes of resolving conflict?" At a conceptual level, the two questions are very similar. Nevertheless, the latter phrasing is much sexier; it resonates directly to many aspects of life and to the other social sciences as well as psychology. And it is also directly connected to many of the social issues with which I was concerned: war and peace, intergroup relations, class conflict, and family conflict.

It was a productive reframing that led to much research in our social psychology laboratory by my students and myself. My book *The Resolution of Conflict: Constructive and Destructive Processes*, published in 1973, summarizes much of this research and had a considerable impact in the social sciences. It helped to provide a new way of thinking about conflict and broadened the focus of the field to include constructive conflicts as well as destructive ones.

Our research into the question central to *The Resolution of Conflict* started off with the assumption that if the parties involved in a conflict situation had a cooperative rather than competitive orientation toward one another, they would be more likely to engage in a constructive process of conflict resolution. In my earlier research on the effects of cooperation and competition upon group process, I had demonstrated that a cooperative process was more productive than a competitive process in dealing with a problem that a group faces. I reasoned that the same would be true in a mixed-motive situation of conflict. A conflict could be viewed as a mutual problem facing the conflicting parties. Our initial research on trust and suspicion employing the Prisoner's Dilemma game strongly supported my reasoning, as did subsequent research employing other experimental formats. I believe that this is a very important result that has considerable theoretical and practical significance.

At a theoretical level, it enabled me to link my prior characterization of cooperation and competitive social processes to the nature of the processes of conflict resolution that would typically give rise to constructive or destructive outcomes. That is, I had found a way to characterize the central features of constructive and destructive *processes* of conflict resolution; doing so represented a major advance beyond the characterization of *outcomes* as constructive and destructive. This not only was important in itself, but it also opened up a new possibility: that we would be able to develop insight into the conditions that initiated or stimulated the development of cooperative-constructive versus competitive-destructive processes of conflict. Much of the research my students and I have done has been addresses to developing this insight.

Much of our early research on the conditions affecting the course of conflict was done on an ad hoc basis. We selected independent variables to manipulate based on our intuitive sense of what would give rise to a cooperative or competitive process. We did experiments with quite a number of variables: motivational orientation, communication facilities, perceived similarity of opinions and beliefs, size of conflict, availability of threats and weapons, power differences, third-party interventions, strategies and tactics of game playing by experimental stooges, the payoff structure of the game, personality characteristics, and so on. The results of these studies fell into a pattern that I slowly began to grasp.

All of these studies seemed explainable by the assumption, which I have labeled "Deutsch's crude law of social relations," that the characteristic processes and effects elicited by a given type of social relationship (cooperative or competitive) also tend to elicit that type of social relationship. *Thus cooperation induces and is induced* by a perceived similarity in beliefs and attitudes, a readiness to be helpful, openness in communication, trusting and friendly attitudes, sensitivity to common interests and de-emphasis of opposed interests, an orientation toward enhancing mutual power rather than power differences, and on. Similarly, competition induces and is induced by the use of tactics of coercion, threat, or deception; attempts to enhance the power differences between oneself and the other; poor communication; minimization of the awareness of similarities in values and increased sensitivity to opposed interests; suspicion and hostile attitudes: the importance, rigidity, and size of the issues in conflict, and so on.

In other words, if one has systematic knowledge of the effects of cooperation and competitive processes, one will have systematic knowledge of the conditions that typically give rise to such processes and, by extension, to the conditions that affect whether a conflict will take a constructive or destructive course. My early theory of cooperation and competition is a theory of the *effects* of cooperative and competitive processes. Hence, from the crude law of social relations stated earlier, it follows that this theory provides insight into the conditions that give rise to cooperative and competitive processes.

The crude law is *crude*. It expresses surface similarities between effects and causes; the basic relationships are genotypical rather than phenotypical. The crude law is crude, but it can be improved. Its improvement requires a linkage with other areas in social psychology, particularly social cognition and social perception. Such a linkage would enable us to view phenotypes in their social environments in such a way as to lead us to perceive correctly the underlying genotypes. We would then be able to know under what conditions "perceived similarity" or "threat" will be experienced as having an underlying genotype different from the one that is usually associated with its phenotype.

Although the gaming conflicts in the laboratory during this period (1963–1973) were relatively benign, the conflicts in the outside world were not. During this period the cold war escalated; the Berlin crisis occurred; the brothers John and Robert Kennedy and Martin Luther King Jr. were assassinated; the United States was increasingly involved in the Vietnam War; there were teach-ins, campus upheavals, race riots, Woodstock, love-ins, communes, the emergence of the new left, and so on. I was not immune to the effects of these events, personally or professionally.

Professionally, as a result of *Preventing World War III* (of which I was coeditor), my activities in SPSSI, my various speeches, and our conflict studies, I became identified as one of the psychologists (along with Ralph White, Charles Osgood, Irving Janis, Jerome Frank, and Herbert Kelman) concerned with war and peace issues. I was invited to participate in meetings on the Berlin crisis, arms control, deterrence, Soviet-U.S. relations, and so on. Some involved high-level diplomats, others involved people in the defense establishment, others were at the U.N., and still others were with citizen groups or social scientists. During the 1960s I was also trying to get more of my fellow psychologists involved in these issues. I took the opportunity of several addresses to speak to these issues: My 1960 SPSSI presidential address was "Psychological Alternatives to War" (Deutsch 1960); my 1966 New York State Psychological Association talk was "Vietnam and the Start of World War III: Some Psychological Parallels" (Deutsch 1966); my 1968 Eastern Psychological Association presentation was "Socially Relevant Science" (Deutsch 1969b); and my Kurt Lewin Memorial Award address was "Conflicts: Productive and Destructive"[4] (Deutsch 1969a).

---

[4] In 1968 I also gave this address at a meeting of social psychologists from the West (the United States and Western Europe) and from Eastern Europe. We met in Prague shortly after the Soviet Union had sent its troops into Czechoslovakia to squash an incipient rebellion against Soviet domination. Despite our misgivings, we came at the strong urging of our Czech colleagues who wanted to maintain their contacts with the West.

My paper included a section on what strategies and tactics were available to "low-power" groups when confronting "high-power" groups. The Czechs loved it and widely circulated a tape recording they made of it.

Leon Festinger, in contrast, asked me, "Is this science?" I replied, "Leon, you and I have a different conception of the nature of science." My conception, I believe, was more inclusive than his. Leon and his followers were always puzzled by me: They thought I did fine theoretical and experimental work, but they did not understand my willingness to apply the best available social science knowledge to important social issues even when that knowledge was not firmly rooted in experimental research.

The meeting in Prague was sponsored by the Transnational Social Psychology Committee of the Social Science Research Council (SSRC). Leon was its chair, and under his leadership it did much to stimulate the development of social psychology in Western Europe.

However, Leon was very much annoyed and harshly criticized Henry Tajfel for his manuscript "Experiments in a Vacuum" and Serge Moscovici for his "Society and Theory in Social

About the time I was finishing the manuscript for my conflict book, in May 1972, I received from Melvin J. Lerner, then at the University of Waterloo, an invitation to participate in a conference entitled "Contributions to a Just Society." Mel had been an NYU social psychology student who had worked with Isadore Chein but had taken some courses with me. Shortly after the conference, he asked me to contribute to the *Journal of Social Issues* volume on the justice motive that he was editing. The two papers I wrote as a result of his urgings were "Awakening the Sense of Injustice" (Deutsch 1974) and "Equity, Equality, and Need: What Determines Which Value Will Be Used as the Basis of Distributive Justice?" (Deutsch 1975). In preparing these papers, I reviewed the existing work on the social psychology of justice and became quite dissatisfied with the dominant approach to this area: equity theory. My dissatisfaction led me to write an extensive critique of equity theory in 1977 (Deutsch 1978, 1979) and, with the support of the National Science Foundation, to embark on a program of research on the social psychology of distribute justice. This program was, without my full recognition, something I had been engaged in for many years. Like Moliere's bourgeois gentleman, I had been "speaking justice" all the time without being aware of it. My dissertation study could be thought of as a study of two different systems of distribute justice, cooperative-egalitarian and competitive-meritocratic. Our research

---

(Footnote 4 continued)
Psychology," both of which were critical of American social psychology. This occurred during a committee meeting in West Germany in 1971.

The committee also exerted some efforts to develop social psychology in Latin America. We held a seminar in Chile for Latin American social psychologists during the tumultuous period just prior to Salvador Allende's coming to power. After Leon resigned as the committee chairman, I was asked to take on this role. We had another East-West meeting in Hungary, in a small resort village about 20 miles from Budapest. We also held a conference in Majorca that led to the book Applying Social Psychology (Deutsch/Hornstein 1975). About this time, SSRC decided to end its financial support for the committee (it had had a rather extended life by SSRC's usual standards for committees). The committee, however, was not quite ready to quit. Martin Irle hosted a small meeting in Mannheim, Germany. I hosted an even smaller one in my beach house in East Hampton, New York, and Jujuji Misumi hosted an even smaller one in Japan.

This traveling committee, which met mainly outside the United States (so as to stimulate the development of social psychology elsewhere), included—at different times—such people as Leon Festinger, John Lanzetta, Stanley Schachter, Harold Kelley, Henry Riecken, and myself from the United States, as well as Serge Moscovici, Henry Tajfel, Jaap Kookebacker, Martin Irle, Ragnar Rommetveit, Jujuji Misumi, and Jaromir Janousek from other parts of the world. Throughout much of its existence, Jerome Singer was the committee's witty and tolerant administrator for SSRC.

During much of the same time, there was another traveling committee funded by the Office of Naval Research, through Luigi Petrullo, which met to discuss research on conflict. About half of its members were from the United States and the other half from Western Europe. Its U.S. members included Harold Kelley, Gerald Shure, John Thibaut, John Lanzetta, Dean Pruitt, and myself. Among the European were Serge Moscovici, Henry Tajfel, Claude Faucheux, Claude Flament, and Josef Nuttin Jr. We met about twice a year, alternating locales between Europoe and the United States. We had many good discussions, excellent wine and food, and formed some lasting friendships. We also did a cross-national experiment and bargaining that has rarely been cited. It was a wonderful period to be a social psychologist.

on bargaining and conflict had direct relevance to a central question in the social psychology of justice, namely, what are the conditions that facilitate the establishment of a stable system of justice among interactants that they will consider to be fair?

Our research program had three main components: (1) experimental studies of the effects of different systems of distributive justice, (2) research into the determinants of the choice of distributive systems, and (3) investigations into the sense of injustice. The theory and research that emanated from this program has been presented mainly in my 1985 book, *Distributive Justice*. I believe it is an important extension of the work I had done on conflict.[5] The book received extremely favorable reviews, but I was disappointed that it did not create as much of the stir as I had hoped, despite some of its interesting ideas and provocative research findings. Possibly this was due to my having included in the book many theoretical papers that had been published earlier.

The year 1982 was particularly outstanding for me. I made two important addresses. In one, my presidential address to the International Society of Political Psychology, I developed the concept of "malignant conflict" and described the processes involved in such conflicts and used this discussion as a basis for analyzing the Cold War between the United States and the Soviet Union (Deutsch 1983a, b, c, 1985). The reaction of the audience was very gratifying. In various follow-ups (e.g., interviews, talks, conferences, pamphlets) it received considerable attention.

The second address was my inaugural lecture as the E.L. Thorndike Professor of Psychology and Education at Teachers College. I admired Thorndike both as a psychologist and as a person (after reading an extensive biography of him), but I felt his views about race reflected the ignorance and bigotry prevalent in his time. In my opening remarks, I expressed my admiration for Thorndike but dissociated myself from his statements about racial and ethnic groups. My address was essentially a review of my work in social psychology. However, in a concluding section, I indicated my intention to help to further develop the educational implications and applications of my work on cooperation and conflict resolution. To this end, I proposed establishing a center at Teachers College that would foster cooperative learning and constructive conflict resolution in the schools. At that time, I vainly hoped that I might be able to induce a former student of mine to direct, administer, and raise funds for such a center; I never liked administrative work or raising funds, even though I had been reasonably successful in doing so during my career. In 1986, with the aid of a small grant from President Michael Timpane

[5] Among the many students who contributed directly to this book were Rebecca Curtis, Michelle Find, Sandra Horowitz, Ivan Lansberg, Brian Maruffi, Louis Medvene, Dolores Mei, Marilyn Seiler, Janice Steil, Bruce Tuchman, Janet Weinglass, William Wench Jr., and Cilio Ziviani. Other students in my work groups on justice who have contributed indirectly to this volume include Lorinda Arella, Adrienne Asch, Susan Boardman, Ellen Brickman, Ellen Fagenson, Martha Gephart, Cheryl Koopman, Jay Kantor, Eric Marcus, Susan Opotow, Jorge da Silva Ribeiro, Rony Rinat, Shula Shichman, and Rachel Solomon.

($9,600), I started the center that I later ambitiously name the *International Center for Cooperation and Conflict Resolution* (ICCCR).

In 1982, I also published a paper, "Interdependence and Psychological Orientation," that integrated several strands in my work. Mike Wish and I (while Mike was on the faculty at Teachers College) did some initial work on characterizing the fundamental dimensions of interpersonal relations. This work grew out of some research that my students and I were doing on marital conflict; we felt it would be useful to go beyond personality descriptions of the individual spouses so that we would be able to characterize the couple as a couple in terms of their relations to one another. Using various data-collection procedures and multidimensional scaling methods, we (Wish et al. 1976) came up with five dimensions: cooperation-competition, power distribution, task-oriented versus social-emotional, formal versus informal, and intensity of the relationship.

Morton Deutsch addressing the International Symposium on Social Psychology in Tokyo (1978). *Source* The author's personal photo collection

Previously, I had done much to characterize the social psychological properties of the first dimension, cooperation-competition. Now I sought to do this for the others. Undoubtedly influenced by the popularity of the cognitive approach, I labeled my first attempt "modes of thought." But this title did not seem to be sufficiently inclusive. It appeared to me evident that cognitive processes differ in types of social relations, and I wanted to sketch the nature of some of these differences. However, I also thought that the psychological differences among the types of social relations were not confined to the cognitive processes: Various motivational and moral dispositions were involved as well.

It had been customary to consider these latter predispositions as more enduring characteristics of the individual and to label them "personality traits" or "character orientations." Since my emphasis is on the situationally influenced nature and, hence temporariness of such predispositions, these labels did not seem fitting either. Thus I settled on the term "psychological orientation" to capture the basic theme of this paper, namely, that people orient themselves differently to different types of social relations and that these orientations reflect and are reflected in various cognitive processes, motivational tendencies, and moral dispositions.

At the time I was not doing research in cognitive social psychology, but I was sympathetic to it for two reasons. First, as someone greatly influenced by the Gestalt psychologists as well as by Lewin and Fritz Heider, I felt perceptual and cognitive processes were very important. Second, I felt it was a healthy reaction to the antimentalist views of B.F. Skinner and his followers, which were quite popular in psychology in the 1960s and 1970s. My sympathies for the cognitive approach possibly unconsciously led me to suppress the significant differences between it and my emphasis on psychological orientations. Psychological orientations involve the cognitive but also motivational and moral orientations. In the 1980s, cognitive social psychologists neglected both the motivational and moral aspects of people's orientations to social relations.

More recently, there has been increasing recognition of the importance of motivation, even belated recognition of the relevance of Lewin's approach, which integrates cognition and motivation. However, psychologists have not yet acknowledged that there is a moral, normative feature to every type of social relation and that any reasonably full characterization of the psychological orientation associated with a social interaction (or its perception) will include the person's moral orientation as well as his or her cognitive and motivational orientation. My work in the area of justice, of course, has helped to sensitize me to the importance of moral norms in social situations. I speculate that the neglect of the moral component of psychological orientation is linked to the fact that the study of justice has not been central in the social psychological research literature. The flurry of interest in equity theory died down in the late 1970s with the decrease of interest in dissonance theory. The dissonance component of equity theory was its most interesting psychological feature.

After publishing *Distributive Justice* in 1985, I sought funding from NSF for a program of basic research related to some of the ideas in my paper "Interdependence and Psychological Orientation." Unfortunately, my proposal was not funded. By this time our NIMH-supported, predoctoral training program was no longer in existence; NIMH's interest had turned toward postdoctoral training. Teachers College provided no funds for research or for graduate research assistants and little secretarial support or money for travel or equipment. It was also a period in which academic appointments became scarce. The consequence was that our doctoral students increasingly became part-time students who often had full-time jobs. In addition, they became more interested in nonacademic positions and more frequently decided to specialize in the organizational rather than in the social psychology component of our doctoral program in social and organizational psychology.

In this context I discontinued my basic research, which had been primarily conducted in the laboratory. From 1985 on, I continued to write and publish papers mainly for small conferences related to conflict or justice, several as award addresses for honors I was receiving and a number by invitation of editors of books or special journal issues. Among the thirty articles I have published since 1985, several titles stand out: "On Negotiating the Non-Negotiable"; "Psychological Consequences of Different Forms of Social Organization"; "The Psychological Roots of Moral Exclusion"; "Sixty Years of Conflict"; "Equality and Economic Efficiency: Is There a Trade-Off?" "Kurt Lewin: The Tough-Minded and Tender-Hearted Scientist"; "Educating for a Peaceful World"; "The Effects of Training in Cooperative Learning and Conflict Resolution in an Alternative High School"; "Constructive Conflict Resolution: Theory, Research, and Practice"; (with Peter Coleman) "The Mediation of Interethnic Conflict"; "William James: The First Peace Psychologist"; and "Constructive Conflict Management for the World Today" (see citations in the References).

1.8 The International Center for Cooperation and Conflict Resolution

In 1986 I started the center that I promised in my Thorndike inaugural address. Our first activity was a workshop to which I invited the superintendents of school districts in and around New York City as well as representatives of several foundations who might become interested in financing the activities for our center. In addition to introductory remarks made by the president of Teachers College and myself, the workshop consisted of a series of miniseminars chosen to reflect the kinds of activities in which our center would engage: cooperative learning, the constructive use of controversy in teaching, conflict resolution training in schools, the training of student mediators, and research evaluation of programs. Each seminar was conducted by a leading expert (e.g., David and Roger Johnson led the seminars on cooperative learning and the constructive use of controversy).

As the result of this workshop, one of the superintendents invited us to develop a program of cooperative learning in his wealthy, suburban school district and to evaluate the program. We sought without success to broaden the program to include conflict resolution training. However, the superintendent was helpful in arranging for us to meet with the superintendent of a nearby, comparable school district that would serve as a control. We approached several foundations for funds but were rejected until I noticed in a publication that Hank Riecken was on the board of the W.T. Grant Foundation. I contacted Hank and told him of our plans and hopes, and he arranged for me to meet with the president and himself. Both were enthusiastic about our plans, which called for support for 5 years at a level of $200,000 per year, and they asked me to write a detailed proposal for submission to the board. The board approved the project for 3 years and indicated that after the first year we should obtain half our funds from other sources. At the time I did not realize that this was a customary but nasty policy of many foundations—forcing one to remain continuously in a fund-raising mode.

We began the project with a preliminary workshop in which David Johnson got a group of senior, influential teachers involved in cooperative learning. They became enthusiastic supporters. Our next step, which proved to be fatal, was to introduce the questionnaires, observational measures, and other recorded data we wished to obtain. We needed permissions from the school board as well as from the school personnel and parents of the students. When the school board learned that we were not only interested in academic achievements but also in measuring social skills, social relations, and psychological adjustment, they were horrified and canceled permission to do the study in their district. As the superintendent regretfully explained, the political attitudes of the board members were to the right of Attila the Hun, and they thought of mental health as a dangerous, explosive topic.

At this point I was sorry that I had left the social psychology laboratory to do research in field settings. However, Ellen Raider, who had joined our center as training director after we were funded, came up with the center-saving suggestion that we move our project to an inner city, alternative high school where she knew the principal and associate principal. Luckily, the foundation was happy to approve the move; they preferred that our research be done with inner-city youth.

I shall not describe the many headaches and heartaches we had in carrying out our research other than to indicate that we were training overworked and fatigued but dedicated teachers, most of whose students lived in poor and difficult circumstances and often did not have the reading or writing skills necessary for successful work as high school students. Also, to put it bluntly, the physical conditions of the school and neighborhood were horrible. Many aspects of the project were not executed as well as we had planned: the training of the teachers; the measurement of the effects on students; the duration of the study; the records kept by the school on student attendance, dropouts, disruptions, and so on. By the standards of a laboratory experiment, it was very unsatisfactory research. Yet I must say that I came out of this study with a great deal of appreciation of those researchers who are foolhardy enough to leave the laboratory. They must have the kind of administrative and social skills, flexibility, ingenuity, statistical wizardry, and frustration tolerance rarely required in laboratory studies.

Despite our problems, much to our surprise, we were able to demonstrate that our training had important and significant effects on the students. In brief, the data showed that as students improved in managing their conflicts (whether or not because of the training in conflict resolution and cooperative learning), they experienced increased social support and less victimization from others. This improvement in their relations with others led to greater self-esteem as well as a fewer feelings of anxiety and depression and more frequent positive feelings of well being. The higher self-esteem, in turn, produced a greater sense of personal control over their own fates. The increases in their sense of personal control and in their positive feelings of well-being led to higher academic performances. There is also indirect evidence that the work readiness and actual work performance of students were also improved. Our data further indicated that students, teachers, and administrators had generally positive views about the training and its results.

This study was the first longitudinal study of the effects of cooperative learning and conflict resolution training conducted in a very difficult school environment. It was also the first to go beyond the measurement of consumer satisfaction. Its positive results were consistent with our theoretical model and with results obtained in smaller, brief studies in experimental classrooms. In part because the study was conducted in the New York City school system, the city's board of education made a contract with ICCCR in 1992–1994. The contract specified that ICCCR would train two key faculty of staff people from every high school in New York City so that one would become sufficiently expert to be able to train students, teachers, and parents in constructive conflict resolution and the other would become sufficiently expert in mediation to be able to establish and administer an effective mediation center at the school, with students functioning as mediators.

Ellen Raider and her staff conducted the training, which took place for 50 h over ten sessions, for a total of 300 people in cohorts over a year and a half. The training methods were based on a model and manuals developed by Ellen Raider and Susan Coleman. The principals of the various high schools also received training in conflict resolution and mediation in 3-day workshops, abbreviated versions of the larger sessions.

Although ICCCR was not provided with funds to conduct a research evaluation of its training, the research division of the board of education and the Dispute Resolution Center of John Jay College were able to conduct some relevant research. The research indicated that within 2 years of training almost all of the more than 150 high schools that participated had established mediation centers in their schools (fewer than 5 % had not). In addition, most of the schools had introduced into their curriculum education in constructive conflict resolution, and thousands of students had exposure to such education. All participants in the research believed that the program had a positive impact on personal relationships and school climate overall. Cited were improvements in the way students dealt with anger and resolved conflicts, heightened respect for differences, better communication skills, and increased understanding of students' needs on the part of the school staff. Some people noted that the school atmosphere was calmer and more collaborative. Peer mediators, disputants, and students who had participated in lessons in cooperative negotiation all commented on positive changes in their own interactions with others, both within and outside of school. Most telling, perhaps, was that disputants had enthusiastically recommended peer mediation to their friends, and curriculum students believed that all students should be required to take lessons in conflict resolution.

ICCCR continues to do conflict resolution training in various school systems and in other contexts, such as the United Nations. More recently, as a prelude to offering graduate studies in conflict resolution at Teachers College, Ellen Raider conducted workshops on conflict resolution with various members of the faculty. The graduate studies now exist as one of the concentrations in the degree programs in social and organizational psychology as well as a certificate program for non-degree students. I have continued to teach a theory course entitled “Fundamentals of Cooperation, Conflict Resolution, and Mediation.” Ellen and her staff have been conducting our various practical courses in this area.

I have also been the organizer for a faculty seminar on conflict resolution from which a book is now in preparation, *The Handbook of Constructive Conflict Resolution: Theory and Practice*, to be published by Jossey-Bass in 2000. I have written four chapters for it, and I am serving as its editor along with Peter Coleman, who is the new Director of ICCCR. As I have reduced my academic responsibilities (less teaching, no more faculty meetings, only one or two highly selected doctoral students whom I supervise), Lydia and I have been doing considerably more traveling and dining in superb restaurants.

1.9 Conclusion

As I look back upon my career, several things stand out for me.

Luck. I was lucky to go to CCNY, which had two young faculty members, Max Hertzman and Walter Scott Neff, who stimulated my interest in Lewin and in social psychological research. I was extremely lucky to be a student at the RCGD at MIT, where I was able to become part of a small, innovative group of faculty and students who had a major impact on the development of modern social psychology. Moreover, my career got off to a quick start largely as a result of the prodding of Stuart Cook, who had me involved in writing two books shortly after I obtained my Ph.D. Also, I was fortunate to be able to receive financial support for my research throughout most of my career. In addition, I have had the opportunity to work with many excellent, productive students who have stimulated me and contributed much to my research. Not least, I was lucky enough to marry a woman whose esthetic sensibility and practical skills helped to create a congenial and supportive home environment that enabled me to focus my attention on scholarly activities rather than on such household activities as fixing things (which I never could do anyway).

Continuing Themes. My work on social psychology has been dominated by two continuing themes with which I have been preoccupied throughout my career. One is my intellectual interest in cooperation and competition, which has been expressed in my theorizing and research in the effects of cooperation and competition, our studies of conflict processes, and our work on distributive justice. I have continued to believe that these foci are central to understanding social life and also that a ‘social’ social psychology rather than an ‘individual’ social psychology would have these as its fundamental concerns. The second continuing interrelated theme has been developing my work so that it has social relevance to key social problems. Sometimes images, derived from such social problems as war and peace, prejudice, marital conflict, and injustice, would be the starting point for the development of a theoretical analysis or an experimental study. At other times I would use theory and research (other social scientists’ as well as mine) in an attempt to shed light in important applications, particularly in the field of education, where I am considered to be one of the parents of cooperative learning and conflict resolution training.

Episodic Research. Occasionally, I strayed from the two themes just described, to do single studies that expressed my reservations about some of the fashionable theorizing and research. I took potshots at Solomon Asch's neglect of group factors in his conformity studies, at Festinger's omission of defensiveness in his dissonance theorizing, at equity theory's assumptions of greater productivity when people are rewarded in proportion to their performance, at social perception studies that ignored the social and institutional context in which social acts are imbedded, and at Henry Tajfel's initial assumption that the mere awareness of a difference among a collection of individuals will promote group formation. My straying was usually short-lived because my primary interests were in the two themes described above and I was not sufficiently energetic to take on additional themes.

Familial Context. As I look back on my career, I am impressed by how much its themes have been influenced by my experiences within my family as well as what was occurring in the broader society. Within my family, I was the youngest of four sons, and I felt a strong need to catch up with my next older brother (two and a half years older), believing that if I did not I would be excluded. In fact, one of my earliest memories focuses on injustice. I was about three and a half years old. We were all staying at a resort in the Catskills, and a counselor organized a game of softball for the older kids (the 6–8-year-olds). I was excluded from it because I was too young and was asked to stay on the side. I was very mad, and when a foul ball was hit near me, I recall picking it up, running with it, and throwing it as far as I could in a direction away from the players. I trace my passionate feeling about injustice to such early experiences as this one. In my attempts to keep up with the older kids, as a child and youth I was quite competitive. However, it was a strain, and when I lost I felt injured and when I won, surpassing my older brother and his friends, I could feel their sense of hurt and shame. My questioning of the value of competition undoubtedly arose from these episodes. This questioning was reinforced by the favorable attitude toward socialism that was held by my parents and many of their friends. My father became rabidly antiunion during my rebellious adolescence, and, perversely, this strengthened my favorable view of unions, cooperation, and socialism.

The Social Context. I grew up in a time when, as a Jew, I experienced many instances of prejudice, blatant as well as subtle, and could observe the gross acts of injustice being suffered by blacks. In my youth and adolescence, there was the economic depression, union organizing, the Spanish civil war, and the emergence of fascism, Nazism, and Stalinism. I was politically engaged—contributing lunch money to the Spanish loyalists, organizing strikes in high school and in a summer resort, participating in a sit-in against the fascist ambassador, and so forth. It is no wonder that I was attracted to Lewin, who I saw as taking psychology in a direction that would enable it to contribute to the development of a democratic cooperative society that was free of prejudice.

The activist theme in my career as a social psychologist undoubtedly reflects the social context of my youth. The social context also helps to explain why I did not become a political activist or union organizer. In my family, among my fellow (mostly Jewish) students, and in my high school and college, there was a strong emphasis on ideas and intellectual achievement. Our heroes were those who contributed to the world through their ideas—Darwin, Marx, Freud, and Einstein. They had exemplified Lewin's dictum, recalled earlier, "there is nothing so practical as a good theory." This has been the second theme of my career.

One final note: Every society has its own implicit assumptions of which its members are usually not aware. We live in a highly individualistic society. Its ethos is that of the lone, self-reliant, enterprising individual who has escaped from the restraints of an oppressive community so as to be free to pursue his or her destiny in an environment that offers ever-expanding opportunity to those who are fittest. I think this image has influenced much of American social psychology, which has been too focused on what goes on in the isolated head of the subject, with a corresponding neglect of the social reality in which the subject is participating.

The socialist ethos incorporates the view that the human being is a social animal whose nature is determined by the way people are related to one another in their productive activities in any given community. Its vision is of social beings free to cooperate with one another toward common objectives because they jointly control the means of production and share the rewards of their collective labor. This vision is a useful supplement to the dominant emphasis in American social psychology. However, it is neglectful of the characteristics of individual persons—characteristics that are determined mainly in the course of interaction between the biological person and his or her social environment.

I conclude with the hope that future social psychologists will be more concerned than we have been with characterizing the socially relevant properties of individuals and the psychologically relevant attributes of social structures. To oversimplify it, I hope that they will provide a successful integration of the orientations of three of the intellectual heroes of my youth: Freud, Marx, and Lewin.

References

Adorno, T.W.; Frenkel-Brunswik, E.; Levinson, D.J.; Sanford, R.N., 1950: *The Authoritarian Personality* (New York: Harper).

Allport, G.W., 1954a: "The Historical Background of Modern Social Psychology", in Lindzey G. (Ed.): *Handbook of Social Psychology*, Volume 1 (Cambridge, MA: Addison-Wesley).

Allport, G.W., 1954b: *The Nature of Prejudice* (Reading, MA: Addison-Wesley).

Bartlett, E.C., 1932: *Remembering: A Study in Social and Experimental Psychology* (Cambridge: Cambridge University Press).

Brown, J.F., 1936: *Psychology and the Social Order: An Introduction to the Dynamic Study of Social Fields* (New York: McGraw-Hill).

Cohen, M.; Nagel, E., 1934: *An Introduction to Logic and Scientific Method* (New York: Harcourt).

Coleman, P.; Deutsch, M., 1995: "The Mediation of Interethnic Conflict", in: Hawley W.D.; Jackson, A.W. (Eds.): *Toward a Common Destiny: Race Relations in America* (Jossey-Bass).

Cooley, C.H., 1956: *Human Nature and the Social Order* (Glencoe, IL: Free Press).

Deutsch, M., 1949a: "An Experimental Study of Effects of Cooperation and Competition upon Group Processes", in: *Human Relations*, 2: 199–231.
Deutsch, M., 1949b: "A Theory of Cooperation and Competition", in: *Human Relations*, 2: 129–151.
Deutsch, M., 1954: "Field Theory in Social Psychology", in: Lindzey G. (Ed.): *Handbook of Social Psychology* (Cambridge, MA: Addison-Wesley): 181–222.
Deutsch, M., 1958: "Trust and Suspicion", in: *Journal of Conflict Resolution*, 2: 265–279.
Deutsch, M., 1960: "Psychological Alternatives to War", in: *Journal of Social Issues*, 18: 97–119.
Deutsch, M., 1961: "The Interpretation of Praise and Criticism as a Function of Their Social Context", in: *Journal of Abnormal and Social Psychology*, 62: 391–400.
Deutsch, M., 1962a: "Cooperation and Trust: Some Theoretical Notes", in: Jones, M. (Ed.): *Nebraska Symposium on Motivation* (Lincoln: University of Nebraska Press): 275–318.
Deutsch, M., 1962b: "Dissonance or Defensiveness?", in: *Journal of Personality*, 30: 16–28.
Deutsch, M., 1966: *Vietnam and the Start of World War III: Some Psychological Parallels* (Presidential Address before the New York State Psychological Association, New York) May 6.
Deutsch, M., 1969a: "Conflicts: Productive and Destructive", in: *Journal of Social Issues*, 25: 7–41.
Deutsch, M., 1969b: "Socially Relevant Science: Reflections and Some Studies of Interpersonal Conflict", in: *American Psychologist*, 24: 1076–1092.
Deutsch, M., 1973: *The Resolution of Conflict: Constructive and Destructive Processes* (New Haven, CT: Yale University Press).
Deutsch, M., 1974: "Awakening the Sense of Injustice", in: Lerner, M.; Ross M. (Eds.): *The Quest for Justice* (New York: Holt).
Deutsch, M., 1976: Equity, Equality, and Need: What Determines Which Value will be used as the Basis for Distributive Justice?", in: *Journal of Social Issues*, 31: 137–150.
Deutsch, M., 1978: "The Social Psychology of Justice", in: *Proceedings of the International Symposium on Social Psychology* (Kyoto, Japan: Japanese Group Dynamics Association): 23–46.
Deutsch, M., 1982: "Interdependence and Psychological Orientation", in: V. Derlega & J.L.
Grzelek (Eds.): Cooperation *and Helping Behavior: Theories and Research* (New York: Academic Press): 15–42.
Deutsch, M., 1983a: "Conflict Resolution: Theory and Practice", in: *Political Psychology*, 4: 431–453.
Deutsch, M., 1983b: "Current Perspectives on Justice", in: *European Journal of Social Psychology*, 13: 305–319.
Deutsch, M., 1983c: "Preventing World War III: A Psychological Perspective", in: *Political Psychology*, 3: 3–31.
Deutsch, M., 1985: *Distributive Justice: A Social Psychological Perspective* (New Haven, CT: Yale University Press).
Deutsch, M., 1988: "*On Negotiating the Non-negotiable*", in: Kellerman, B.; Rubin J. (Eds.): *Leadership and Negotiation in the Middle East* (New York: Praeger): 248–263.
Deutsch, M., 1989: "Equality and Economic Efficiency: Is There a Trade-Off?", in: Eisenberg, N.; Reykowski, J.; Staub, E. (Eds.): *Social and Moral Values* (Hillsdale, NJ: Erlbaum).
Deutsch, M., 1990a: "Psychological Consequences of Different Forms of Social Organization", in: Himmelweit, H.; Gaskell, G. (Eds.): *Societal Psychology* (London: Sage): 157–178.
Deutsch, M., 1990b: "Sixty Years of Conflict", in: *International Journal of Conflict Management*, I: 237–263.
Deutsch, M., 1991: "Egalitarianism in the Laboratory and in the Workplace", in: Vermunt, R.; Steensma, H. (Eds.): *Social Justice in Human Relations*, Volume 1 (New York: Plenum Press): 195–210.
Deutsch, M., 1992a: *The Effects of Training in Cooperative Learning and Conflict Resolution in an Alternative High School* (New York: International Center for Cooperation and Conflict Resolution, Teachers College, Columbia University).
Deutsch, M., 1992b: Kurt Lewin: "The Tough-Minded and Tender-Hearted Scientist", in: *Journal of Social Issues*, 48: 31–43.
Deutsch, M., 1993a: "Educating for a Peaceful World", in: *American Psychologist*, 48: 510–517.
Deutsch, M., 1993b: "The Effects of Training in Cooperative Learning and Conflict Resolution in an Alternative High School", in: *Cooperative Learning*, 13: 2–5.

Deutsch, M., 1994a: "Constructive Conflict Management for the World Today", in: *International Journal of Conflict Management*, 5: 111–129.

Deutsch, M., 1994b: "Constructive Conflict Resolution: Theory, Research, and Practice", in: *Journal of Social Issues*, 50: 17–26.

Deutsch, M., 1995: "William James: The First Peace Psychologist", in: *Peace and Conflict: Journal of Peace Psychology*, 1: 17–26.

Deutsch, M.; Collins, M.E., 1951: "Interracial Housing: A Psychological Evaluation of a Social Experiment", (Minneapolis: University of Minnesota Press).

Deutsch, M.; Gerard, H.B., 1955: "A Study of Normative and Informational Influences upon Individual Judgment", in: *Journal of Abnormal and Social Psychology*, 51, 629–636.

Deutsch, M.; Hornstein, A. (Eds.) 1975: *Applying Social Psychology: Implications for Research, Practice and Training* (Hillsdale, NJ: Erlbaum).

Deutsch, M.; Krauss, R., 1960: "The Effect of Threat upon Interpersonal Bargaining", in: *Journal of Abnormal and Social Psychology*, 61: 181–189.

Deutsch, M., Krauss, R., 1962: "Studies of Interpersonal Bargaining", in: *Journal of Conflict Resolution*, 6: 52–76.

Deutsch, M.; Krauss, R., 1965: *Theories in Social Psychology* (New York: Basic Books).

Johnson, D.W.; Johnson, R.T., 1989: *Cooperation and Competition: Theory and Research* (Edina, MN: Interaction).

Lewin, K., 1935: *A Dynamic Theory of Personality* (New York: McGraw-Hill).

Lewin, K., 1936: *Principles of Topological Psychology* (New York: McGraw-Hill).

Lewin, K., 1938: *The Conceptual Representation and Measurement of Psychological Forces* (Durham, NC: Duke University Press).

Lewin, K., 1951: *Field Theory in Social Science* (New York: Harper).

Marrow, A., 1969: *The Practical Theorist: The Life and Work of Kurt Lewin* (New York: Basic Books).

McDougall, W., 1908: *An Introduction to Social Psychology* (London: Methuen).

Piaget, J., 1948: *The Moral Judgement of the Child* (Glencoe, IL: Free Press).

Ross, E.A., 1908: *Social Psychology* (New York: Macmillan).

Selltiz, D.; Jahoda, M.; Deutsch, M.; Cook, S.W., 1959: *Research Methods in Social Relations*, 2nd edition (New York: Hold and Dryden).

Sherif, M., 1936: *The Psychology of Social Norms* (New York: Harper).

Tarde, G., 1903: *The Laws of Imitation* (New York: Henry Holt).

Watson, J.B., 1930: *Behaviorism*, Revised edition (New York: W.W. Norton).

Wish, M.; Deutsch, M.; Kaplan, S., 1976: "Perceived Dimensions of Interpersonal Relations", in: *Journal of Personality and Social Psychology*, 33: 409–420.

Wright, Q.; Evan, W.M.; Deutsch, M. (Eds.) 1962: *Preventing World War III: Some Proposals* (New York: Simon and Schuster).

Chapter 2
Selected Bibliography of Morton Deutsch's Writings

2.1 Books in Print

Collins, M.E., 1951: *Interracial Housing: A Psychological Evaluation of a Social Experiment* (Minneapolis: University of Minnesota Press).[1]
Wright, Q.; Evan, W.M., 1962: *Preventing World War III: Some Proposals* (New York: Simon and Schuster).
1985: *The Resolution of Conflict: Constructive and Destructive Processes* (New Haven: Yale University Press).
Hornstein, H., 1975: *Applying Social Psychology: Implications for Research, Practice and Training* (Hillsdale, NJ: L. Erlbaum Associates).
1985: *Distributive Justice: A Social Psychological Perspective* (New Haven: Yale University Press).
Coleman, P.T.; Marcus, Eric, 2014: *The Handbook of Conflict Resolution: Theory and Practice* (San Francisco: Jossey-Bass).
Coleman, P.T., 2012: *Psychological Components of Sustainable Peace* (New York: Springer).

[1] A comprehensive bibliography of Morton Deutsch's writings is available for download at: http://afes-press-books.de/html/SpringerBriefs_PSP_MortonDeutsch.htm.

P.T. Coleman and M. Deutsch, *Morton Deutsch: A Pioneer in Developing Peace Psychology*, SpringerBriefs on Pioneers in Science and Practice 30,
DOI 10.1007/978-3-319-15440-4_2

Ed Simpson and Morton Deutsch, 1984. *Source* The author's Personal Photo collection

2.2 Journal Articles and Chapters in Books

1949: "A Theory of Cooperation and Competition" in: *Human Relations*, 2: 129–151.

1949: "An Experimental Study of Effects of Cooperation and Competition Upon Group Processes", in: *Human Relations*, 2: 199–231.

1961: "Some Considerations Relevant to National Policy", in: *Journal of Social Issues*, 17: 57–68.

1962: "A Psychological Basis for Peace", in: Quincy Wright; Evan, W.M.; Morton Deutsch (Eds.): *Preventing World War III: Some Proposals: Some Proposals* (New York: Simon and Schuster).

1962: "Psychological Alternatives to War", in: *Journal of Social Issues*, 18: 97–119.

1964: "On Producing Change in an Adversary", in: Ronald Fisher (Ed.): *International Conflict and the Behavioral Sciences* (New York: Basic Books): 145–160.

1965: "A Psychological Approach to International Conflict", in: Sperazzo, G. (Ed.): *Strategic Interaction and Conflict* (Washington, D.C.: Georgetown University Press).

1966: "Vietnam and the Start of World War III: Some Psychological Parallels", in: *New York State Psychologist.*

1969: "Conflicts: Productive and Destructive", in: *Journal of Social Issues*, 25: 7–41.

Kelley, H.H.; Lanzetta, J.T.; Nuttin, J., Jr.; Shure, G.H.; Faucheux, C.; Moscovici, S.; Rabie, J., 1970: "A Comparative Experimental Study of Negotiation Behavior", in: *Journal of Personality and Social Psychology*, 16: 411–438.

Wish, M.; Biener, L., 1970: "Differences in Conceptual Structures of Nations: An Exploratory Study", in: *Journal of Personality and Social Psychology*, 16: 361–373.

Wish, M.; Biener, L., 1972: "Differences in Perceived Similarity of Nations", in: Romney, A.K.; Shepherd, R.N. (Eds.): *Multidimensional Scaling in the Social Sciences* (Stanford, CA: Stanford University Press).

1976: "On Cursing the Darkness Versus Lighting a Candle", in: *Social Psychology in Transition* (New York: Plenum).

1980: "Fifty Years of Conflict," in: Festinger, L. (Ed.): *Retrospections on Social Psychology* (New York: Oxford University Press): 46–77.

1982: "Interdependence and Psychological Orientation", in: Derlega, V.; Grzelek, J.L. (Eds.): *Cooperation and Helping Behavior: Theories and Research* (City: Academic Press): 15–42.

1983: "Preventing World War III: A Psychological Perspective", in: *Political Psychology*, 3: 3–31.

1983: "What is Political Psychology?", in: *International Social Science Journal*, 35: 221–236.

Allen, Janet, 1984: "Averting World War III: An Interview with Morton Deutsch", in: *New Ideas in Psychology*: 189–199.

Smith, Brewster, 1984: "Preventing Armageddon: A Social Issue Release" (Washington, D.C.: American Psychological Association).

1986: "Folie a deaux": A Psychological Perspective on Soviet-American Relations", in: Karns, M.P. (Ed.): *Persistent Patterns and Emergent Structures in a Waning Century* (New York: Praeger): 185–196.

1986: "The Malignant (Spiral) Process of Hostile Interaction", in: White, R. (Ed.): *Psychology and the Prevention of Nuclear War* (New York: New York University Press): 131–154.

1986: "Strategies of Inducing Cooperation", in: White, R. (Ed.): *Psychology and the Prevention of Nuclear War* (New York: New York University Press): 162–170.

1987: "Going Beyond 'Beyond Deterrence", in: *Journal of Social Issues*, 43: 143–153.

Steil, J.M., 1988: "Awakening the Sense of Injustice", in: *Social Justice Research*, 2: 3–23.

1988: "Commentary: On Negotiating the Non-negotiable", in: Kellerman, B; Rubin, J. (Eds.): *Leadership and Negotiation in the Middle East* (New York: Praeger): 248–263.

Deutsch and Family: Lydia, Anthony, and Nicholas, 1982: *Source* The author's personal photo collection

"Peace Education: Cooperative Learning, Conflict Resolution Training, and the use of Constructive Controversy," *Reprints and Miniprints* (Malmö, Sweden: School of Education, No. 671, 1989).
1990: "Sixty Years of Conflict", in: *The International Journal of Conflict Management*, 1: 237–263.
1992: "Educating Beyond Hate", in: Bjerstedt, A. (Ed.): *Peace, Environment, and Education*, 2: 3–19.
1993: "Educating for a Peaceful World", in: *American Psychologist*, 48: 510–517.
"Constructive Conflict Management for the World Today", in: *International Journal of Conflict Management*, 5 (1994): 111–129.
James, William, 1995: "The First Peace Psychologist", in: *Peace and Conflict Journal of Peace Psychology*, 1: 17–26.
1999: "A Personal Perspective on the Development of Social Psychology in the 20th Century", in: Rodrigues, A.; Levine, R. (Eds.): *Reflections on 100 Years of Experimental Social Psychology* (Boulder; Co: Westview Press).

Kinnvall, C., 2002: "What is Political Psychology?", in: K. Monroe (Ed.): *Political Psychology* (Mahwah, NJ: Erlbaum).
"Mediation & Difficult Conflicts", in: M. Herrman (Ed.): *Handbook of Mediation* (Malden, MA: Blackwell Publishing): 355–373.
2006: "The Interplay Between Internal and External Conflict", in: Schnieder, A.; Honeyman, C. (Eds.): *The Negotiator's Fieldbook* (Washington, DC: American Bar Association).
2012: "A Theory of Cooperation-Competition and Beyond", in: Van Lange, P.M.; Kruglanski, A.W.; Higgins, E.T. (Eds.): *Handbook of Theories of Social Psychology*, Vol 2 (London–Thousand Oaks, CA: Sage): 275–294.
Marcus, E.C.; Brazaitis, S., 2015: "Developing a Global Community: A Social Psychology Perspective", in: Mauro Galluccio (Ed.): *Handbook of International Negotiation—Interpersonal, Intercultural, and Diplomatic Perspectives* (New York: Springer).

Mort and Harold Kelley at a conference in Kyoto, Japan in October 1978 in the garden of Kyoto International Conference Hall: *Source* The author's personal photo collection

Part II
Texts by Morton Deutsch on Peace Psychology

Mort and Lydia at the Asiate Restaurant with two distinguished former students, Michelle Fine and Susan Opotow. *Source* The author's personal photo collection

Mort giving the E.L. Thorndike Professorship address in receiving this honor at Teachers College, Columbia, 1982. *Source* The author's personal photo collection

Chapter 3
Cooperation, Competition, and Conflict

Some time ago in the garden of a friend's house, my 5-year-old son and his chum were struggling over possession of a water hose.[1] (They were in conflict.) Each wanted to use it first to water the garden. (They had a competitive orientation.) Each was trying to tug it away from the other, and both were crying. Each was very frustrated, and neither was able to use the hose to sprinkle the flowers as he had desired. After reaching a deadlock in this tug-of-war, they began to punch one another and call each other names. As a result of their competitive approach, the conflict took a destructive course for both of them—producing frustration, crying, and violence.

Now imagine a different scenario. The garden consists mainly of two sections, flowers and vegetables. Each kid wants to use the hose first. Let's suppose they want to resolve their conflict amicably. (They have a cooperative orientation.) One says to the other, "Let's flip a coin to see who uses the hose first." (It is a fair procedure for resolving the conflict.) The other agrees and suggests that the loser be given the right to select which section of the garden he waters. They both agree to the suggestion. (They reach a cooperative, win-win agreement.) Their agreements are implemented, and both children feel happy and good about one another. (These are common effects of a cooperative or constructive approach to a conflict.)

As this example illustrates, whether the participants in a conflict have a cooperative orientation or a competitive one is decisive in determining its course and outcomes. This chapter is concerned with understanding the processes involved in cooperation and competition, their effects, and the factors that contribute to developing a cooperative or competitive relationship. It is important to understand the nature of cooperation and competition because almost all conflicts are mixed motive, containing elements of both cooperation and competition.

The theory being presented here was initially developed by Deutsch (1949a, b, 1973, 1985, 2011) and much elaborated by David W. Johnson (Johnson and Johnson 2005, 2011). The Johnsons have provided the most extensive summary of

[1] This text was first published as: "Cooperation, Competition, and Conflict", in: Peter T. Coleman, Morton Deutsch and Eric C. Marcus (Eds.) *The Handbook of Conflict Resolution: Theory and Practice* (3rd Edition). San Francisco: Jossey-Bass, 2014). Permission to republish this text was granted by Brian Collins, Permissions Assistant Wiley, Chichester, UK on 12 November 2014.

P.T. Coleman and M. Deutsch, *Morton Deutsch: A Pioneer in Developing Peace Psychology*, SpringerBriefs on Pioneers in Science and Practice 30,
DOI 10.1007/978-3-319-15440-4_3

the theory and the research bearing on it; their 2005 book and 2011 publication should be consulted for greater detail.

3.1 The Theory of Cooperation and Competition

The theory has two basic ideas. One relates to the type of interdependence among goals of the people involved in a given situation. The other pertains to the type of action that the people involved take.

I identify two basic types of goal interdependence: positive (where the goals are linked in such a way that the amount or probability of a person's goal attainment is positively correlated with the amount or probability of another obtaining his or her goal) and negative (where the goals are linked in such a way that the amount or probability of goal attainment is negatively correlated with the amount or probability of the other's goal attainment). To put it colloquially, if you are positively linked with another, then you sink or swim together; with negative linkage, if the other sinks, you swim, and if the other swims, you sink.

Few situations are purely positive or negative. In most situations, people have a mixture of goals so that it is common for some of their goals initially to be positive and some negatively interdependent. For analytical purposes, I discuss pure situations in this section. In mixed situations, the relative strengths of the two types of goal interdependency, as well as their general orientation to one another, largely determine the nature of the conflict process.

I also characterize two basic types of action by an individual: effective actions, which improve the actor's chances of obtaining a goal, and bungling actions, which worsen the actor's chances of obtaining the goal. (For the purpose of simplicity, I use dichotomies for my basic concepts; the dichotomous types of interdependence and the dichotomous types of actions are, I assume, polar ends of continua.) I then combine types of interdependence and types of action to posit how they jointly affect three basic social psychological processes that I discuss later in this chapter: substitutability, attitudes (cathexis), and inducibility.

People's goals may be linked for various reasons. Thus, positive interdependence can result from people liking one another, being rewarded in terms of their joint achievement, needing to share a resource or overcome an obstacle together, holding common membership or identification with a group whose fate is important to them, being unable to achieve their task goals unless they divide up the work, being influenced by personality and cultural orientation, being bound together because they are treated this way by a common enemy or an authority, and so on. Similarly, with regard to negative interdependence, it can result from people disliking one another or from their being rewarded in such a way that the more the other gets of the reward, the less one gets, and so on.

In addition to positive and negative interdependence, there can be a lack of interdependence, or independence, such that the activities and fate of the people involved do not affect one another directly or indirectly. If they are completely independent of one another, no conflict arises; the existence of a conflict implies some form of interdependence.

One further point: asymmetries may exist with regard to the degree of interdependence in a relationship. Suppose that what you do or what happens to you may have a considerable effect on me, but what I do or what happens to me may have little impact on you. I am more dependent on you than you are on me. In the extreme case, you may be completely independent of me and I may be highly dependent on you. As a consequence of this asymmetry, you have greater power and influence in the relationship than I do. This power may be general if the asymmetry exists in many situations, or it may be situation specific if the asymmetry occurs only in a particular situation. A master has general power over a slave, while an auto mechanic repairing my car's electrical system has situation-specific power.

The three concepts of substitutability, attitudes, and inducibility are vital to understanding the social and psychological processes involved in creating the major effects of cooperation and competition. *Substitutability* (how a person's actions can satisfy another person's intentions) is central to the functioning of all social institutions (the family, industry, schools), the division of labor, and role specialization. Unless the activities of other people can substitute for yours, you are like a person stranded on a desert island alone: you have to build your own house, find or produce your own food, protect yourself from harmful animals, treat your ailments and illnesses, educate yourself about the nature of your new environment and about how to do all these tasks, and so on, without the help of others. Being alone, you can neither create children nor have a family. Substitutability permits you to accept the activities of others in fulfilling your needs. Negative substitutability involves active rejection and effort to counteract the effects of another's activities.

Attitudes refer to the predisposition to respond evaluatively, favorably or unfavorably, to aspects of one's environment or self. Through natural selection, evolution has ensured that all living creatures have the capacity to respond positively to stimuli that are beneficial to them and negatively to those that are harmful. They are attracted to, approach, receive, ingest, like, enhance, and otherwise act positively toward beneficial objects, events, or other creatures. In contrast, they are repelled by harmful objects and circumstances and avoid, eject, attack, dislike, negate, and otherwise act negatively toward them. This inborn tendency to act positively toward the beneficial and negatively toward the harmful is the foundation on which the human potentials for cooperation and love, as well as for competition and hate, develop. The basic psychological orientation of cooperation implies the positive attitude that "we are for each other," "we benefit one another"; competition, by contrast, implies the negative attitude that "we are against one another" and, in its extreme form, "you are out to harm me."

Inducibility refers to the readiness to accept another's influence to do what he or she wants. Negative inducibility refers to the readiness to reject or obstruct fulfillment of what the other wants. The complement of substitutability is inducibility: you are willing to be helpful to another whose actions are helpful to you but not to someone whose actions are harmful. In fact, you reject any request to help the other engage in harmful actions and, if possible, obstruct or interfere with these actions if they occur.

3.2 The Effects of Cooperation and Competition

The theory predicts that if you are in a positively interdependent relationship with someone who bungles, the bungling is not a substitute for effective actions you intended; thus, you view the bungling negatively. In fact, when your net-playing tennis partner in a doubles game allows an easy shot to get past him or her, you have to extend yourself to prevent being harmed by the error. But if your relationship is one of negative interdependence, and the other person bungles (as when your tennis opponent double faults), your opponent's bungle substitutes for an effective action on your part, and you regard it positively. The reverse is true for effective actions. An opponent's effective actions are not substitutable for yours and are negatively valued; a teammate can induce you to help him or her make an effective action, but you are likely to try to prevent or obstruct a bungling action by your teammate. In contrast, you are willing to help an opponent bungle, but your opponent is not likely to induce you to help him or her make an effective action (which, in effect, harms your chances of obtaining your goal).

The theory of cooperation and competition then goes on to make further predictions about different aspects of intrapersonal, interpersonal, intragroup, and intergroup processes from the predictions about substitutability, attitudes, and inducibility. Thus, assuming that the individual actions in a group are much more frequently effective than bungling, among the predictions that follow from the theory are that cooperative relations (those in which the goals of the parties involved are predominantly positively interdependent), as compared with competitive ones, show more of these positive characteristics:

1. *Effective communication is exhibited.* Ideas are verbalized, and group members are attentive to one another, accepting of the ideas of other members and are influenced by them. They have fewer difficulties in communicating with or understanding others.
2. *Friendliness, helpfulness, trust, and lessened obstructiveness* are expressed in the discussions. Members also are more satisfied with the group and its solutions and favorably impressed by the contributions of the other group members. In addition, members of the cooperative groups rate themselves high in desire to win the respect of their colleagues and in obligation to the other members.
3. *Coordination of effort, division of labor, orientation to task achievement, orderliness in discussion, and high productivity* are manifested in the cooperative groups (if the group task requires effective communication, coordination of effort, division of labor, or sharing of resources).

4. Feeling of agreement with the ideas of others and a sense of basic similarity in beliefs and values, as well as confidence in one's own ideas and in the value that other members attach to those ideas, *are obtained in the cooperative groups.*
5. Recognizing and respecting the other by being responsive to the other's needs.
6. *Willingness to enhance the other's power* (e.g., the knowledge, skills, resources, and so on) to accomplish the other's goals increases. As the other's capabilities are strengthened, you are strengthened; they are of value to you as well as to the other. Similarly, the other is enhanced from your enhancement and benefits from your growing capabilities and power.
7. *Defining conflicting interests as a mutual problem to be solved by collaborative effort* facilitates recognizing the legitimacy of each other's interests and the need to search for a solution responsive to the needs of all. It tends to limit rather than expand the scope of conflicting interests. Attempts to influence the other tend to be confined to processes of persuasion.

In contrast, a competitive process has the opposite effects:

1. *Communication is impaired* as the conflicting parties seek to gain advantage by misleading the other through use of false promises, ingratiation tactics, and disinformation. It is reduced and considered futile as they recognize they cannot trust one another's communications to be honest or informative.
2. Obstructiveness and lack of helpfulness lead to mutual negative attitudes, distrust, and suspicion of one another's intentions. *One's perceptions of the other tend to focus on the person's negative qualities and ignore the positive.*
3. *The parties to the process are unable to divide their work*, duplicating one another's efforts such that they become mirror images. If they do divide the work, they feel the need to check continuously what the other is doing.
4. The repeated experience of disagreement and critical rejection of ideas *reduces confidence in oneself as well as the other.*
5. The conflicting parties seek to enhance their own power and reduce the power of the other. *Any increase in the power of the other is seen as threatening to oneself.*
6. *The competitive process stimulates the view that the solution of a conflict can be imposed only by one side on the other*, which leads to using coercive tactics such as psychological and physical threats and violence. It tends to expand the scope of the issues in conflict as each side seeks superiority in power and legitimacy. The conflict becomes a power struggle or a matter of moral principle and is no longer confined to a specific issue at a given time and place. Escalating the conflict increases its motivational significance to the participants and may make a limited defeat less acceptable and more humiliating than a mutual disaster.

As the conflict escalates, it perpetuates itself by such processes as autistic hostility, self-fulfilling prophecies, and unwitting commitments.

Autistic hostility involves breaking off contact and communication with the other; the result is that the hostility is perpetuated because one has no opportunity to learn that it may be based on misunderstandings or misjudgments or to learn if the other has changed for the better.

Self-fulfilling prophecies are those wherein you engage in hostile behavior toward another because of a false assumption that the other has done or is preparing to do something harmful to you; your false assumption comes true when it leads you to engage in hostile behavior that then provokes the other to react in a hostile manner to you. The dynamics of an escalating, destructive conflict have the inherent quality of a *folie a deux* in which the self-fulfilling prophecies of each side mutually reinforce one another. As a result, both sides are right to think that the other is provocative, untrustworthy, and malevolent. Each side, however, tends to be blind to how it and the other have contributed to this malignant process.

In the case of *unwitting commitments*, the parties not only overcommit to rigid positions during the course of escalating conflict but also may unwittingly commit to negative attitudes, perceptions, beliefs, defenses against the other's expected attacks, and investments involved in carrying out their conflictual activities. Thus, during an escalated conflict, a person (a group, a nation) may commit to the view that the other is an evil enemy, the belief that the other is out to take advantage of oneself (one's group, nation), the conviction that one has to be constantly vigilant and ready to defend against the danger the other poses to one's vital interests, and also invest in the means of defending oneself as well as attacking the other. After a protracted conflict, it is hard to give up a grudge, to disarm without feeling vulnerable, as well as to give up the emotional charge associated with being mobilized and vigilant in relation to the conflict.

As Johnson and Johnson (2005, 2011) have detailed, these ideas have given rise to a large number of research studies indicating that a cooperative process (as compared to a competitive one) leads to greater group productivity, more favorable interpersonal relations, better psychological health, and higher self-esteem. Research has also shown that more constructive resolution of conflict results from cooperative as opposed to competitive processes.

For understanding the nature of the processes involved in conflict, this last research finding is of central theoretical and practical significance. It suggests that constructive processes of conflict resolution are similar to cooperative processes of problem solving, and destructive processes of conflict resolution are similar to competitive processes. Because our prior theoretical and research work gave us considerable knowledge about the nature of the processes involved in cooperation and competition, it is evident that this knowledge provides detailed insight into the nature of the processes entailed in constructive and destructive conflict resolution. This kind of knowledge contributes to understanding what processes are involved in producing good or bad outcomes of conflict. There are many ways of characterizing the outcomes of a conflict: the satisfaction or dissatisfaction of the parties, material benefits and costs, improvement or worsening of their relationship, effects on self-esteem and reputation, precedents set, kinds of lessons learned, effects on third parties (such as children of divorcing parents), and so on. Thus, there is reason to believe that a cooperative-constructive process of conflict resolution leads to such good outcomes as mutual benefits and satisfaction, strengthening relationships, positive psychological effects, and so on, while a competitive- destructive process leads to material losses and dissatisfaction, worsening relationships, and negative psychological effects in at least one party (the loser if it is a win-lose outcome) or both parties (if it is a lose-lose outcome).

3.3 Constructive and Destructive Competition

Competition can vary from destructive to constructive: unfair, unregulated competition at the destructive end; fair, regulated competition in between; and constructive competition at the positive end. In constructive competition, the losers as well as the winners gain. Thus, in a tennis match that takes the form of constructive competition, the winner suggests how the loser can improve, offers an opportunity for the loser to learn and practice skills, and makes the match an enjoyable or worthwhile experience for the loser. In constructive competition, winners see to it that losers are better off, or at least not worse off than they were before the competition.

The major difference, for example, between constructive controversy and competitive debate, is that in the former, people discuss their differences with the objective of clarifying them and attempting to find a solution that integrates the best thoughts that emerge during the discussion, no matter who articulates them (see Chap. 4 for a fuller discussion). There is no winner and no loser; both win if, during the controversy, each party comes to deeper insights and enriched views of the matter that is initially in controversy. Constructive controversy is a process for constructively coping with the inevitable differences that people bring to cooperative interaction because it uses differences in understanding, perspective, knowledge, and worldview as valued resources. By contrast, in competitive contests or debates, there is usually a winner and a loser. The party judged to have "the best"—ideas, skills, knowledge, and so on—typically wins, while the other, who is judged to be less good, typically loses. Competition evaluates and ranks people based on their capacity for a particular task rather than integrating various contributions.

By my emphasis throughout this chapter, I do not mean to suggest that competition produces no benefits. Competition is part of everyday life. Acquiring the skills necessary to compete effectively can be of considerable value. Moreover, competition in a cooperative, playful context can be fun. It enables one to enact and experience, in a no serious setting, symbolic emotional dramas relating to victory and defeat, life and death, power and helplessness, dominance and submission—dramas that have deep personal and cultural roots. In addition, competition is a useful social mechanism for selecting those who are better able to perform the activities involved in the competition. Furthermore, when no objective, criterion-referenced basis for measurement of performance exists, the relative performance of students affords a crude yardstick. Nevertheless, serious problems are associated with competition when it does not occur in a cooperative context and if it is not effectively regulated by fair rules. (See Deutsch 1973: 377–388, for a discussion of regulating competition.)

Fair competition is an essential ingredient of a democratic governance process as well as of an effective free market economic system. The democratic process of elections is undermined if the rules and procedures make it more difficult for those who favor a particular party or candidate to vote or have their vote counted). Similarly, if bribery or political influence allows one company or industry to avoid following regulations that others are required to implement, economic efficiency and the free market are undermined.

3.4 Pathologies of Cooperation

As I have indicated in my writings on cooperation and competition (Deutsch 1949a, b, 1973), there is a natural tendency for cooperation to break down as a result of the very social psychological processes—substitutability, attitudes (cathexis), and inducibility—that are central to cooperation. Thus, *substitutability*, which enables the work of one cooperator to replace the work of another without duplicating one another's efforts, leads to function specialization. Specialization of function in turn gives rise to specialized interests and specialized terminology and language; the likely consequence is a deterioration of group unity as those with special interests compete for scarce resources and communicate in a language that is not fully shared. Similarly, *cathexis* of other group members (the development of personal bonds between members) can lead to in-group favoritism, clique formation, nepotism, and so on. Here, the consequences are likely to be a weakening of overall group cohesion as cliques develop, a deterioration of cooperation with other groups as in-group favoritism grows, and a lessening of group effectiveness as a result of nepotism. *Inducibility*, the readiness to be influenced positively by other group members, can lead to excessive conformity with the views of others so that one no longer makes one's own independent, unique contribution to the group. The cooperative process, as a result, may be deprived of the creative contributions made by each of its members; also, those who suppress their individuality may feel inwardly alienated from themselves and their group despite their outer conformity. In addition, free riding or social loafing may occur in which some members shirk their responsibilities to the group and seek to obtain the benefits of group membership without offering the contributions they are able to make to it.

Among the procedures that are employed to prevent the impairment of cooperation:

- Rotation among positions and job enlargement to retard the development of specialized interests.
- Fostering communication among individuals and groups with different interests to facilitate perception of common interests.
- Educating and indoctrinating members so that they become group oriented.

- Developing group symbols, rituals, and occasions to foster group unity and personal identification with the group.
- Instituting coordinating and translating mechanisms, as well as crosscutting memberships in specialized subgroups.
- Honoring and cherishing individuality and buttressing the right to differ.
- Maintaining sufficient individual accountability so that shirking can be detected and responded to with appropriate diagnostic and intervention procedures.
- Engaging in periodic, independent reviews of the way the cooperative system is functioning and making the necessary repairs.

The effort in maintaining effective cooperative systems and repairing them when required is considerable. When cooperation is not required and individual action is feasible, the costs of cooperation may outweigh its benefits and make the individual action preferable. However, often individual action is insufficient and cooperation is necessary. In such cases, the effort required to develop and maintain an effective cooperative process may be the only sensible alternative to the dismal consequences of failure to do so.

3.5 Initiating Cooperation and Competition

If we know that cooperative and competitive processes have important effects on conflict resolution, a question follows: What initiates or gives rise to one or the other process? I did much research (Deutsch 1973) in an attempt to find the answer. The results of my many studies fell into a pattern I slowly began to grasp. They seemed explainable by an assumption I have immodestly labeled "Deutsch's Crude Law of Social Relations":

> The characteristic processes and effects elicited by a given type of social relationship also tend to elicit that type of social relationship, and a typical effect tends to induce the other typical effects of that relationship.

Thus, cooperation induces and is induced by perceived similarity in beliefs and attitudes, readiness to be helpful, openness in communication, trusting and friendly attitudes, sensitivity to common interests and de-emphasis of opposed interests, orientation toward enhancing mutual power rather than power differences, and so on. Similarly, competition induces and is induced by the use of the tactics of coercion, threat, or deception; attempts to enhance the power differences between oneself and the other; poor communication; minimization of the awareness of similarities in values and increased sensitivity to opposed interests; suspicious and hostile attitudes; the importance, rigidity, and size of issues in conflict; and so on.

In other words, someone who has systematic knowledge of the effects of cooperative and competitive processes has systematic knowledge of the conditions that typically give rise to such processes and, by extension, the conditions that affect whether a conflict takes a constructive or destructive course. My early theory of cooperation and competition is a theory of the effects of cooperative and competitive processes. Hence, from the Crude Law of Social Relations, it follows that this theory brings insight into the conditions that give rise to cooperative and competitive processes.

This law is certainly crude. It expresses surface similarities between effects and causes; the basic relationships are genotypical rather than phenotypical. The surface effects of cooperation and competition are due to the underlying type of interdependence (positive or negative) and type of action (effective or bungling), the basic social psychological processes involved in the theory (substitutability, attitudes, and inducibility), and the cultural or social medium and situational context in which these processes are expressed. Thus, how a positive attitude is expressed in an effective, positively interdependent relationship depends on what is appropriate to the cultural or social medium and situational context; that is, presumably one would not seek to express it in a way that is humiliating or embarrassing or likely to be experienced negatively by one's partner.

Similarly, the effectiveness of any typical effect of cooperation or competition as an initiating or inducing condition of a cooperative or competitive process is not due to its phenotype but rather to the inferred genotype of the type of interdependence and type of action. Thus, in most social media and social contexts, perceived similarity in basic values is highly suggestive of the possibility of a positive linkage between oneself and the other. However, we are likely to see ourselves as negatively linked in a context that leads each of us to recognize that similarities in values impel seeking something that is in scarce supply and available for only one of us. Also, it is evident that although threats are mostly perceived in a way that suggest a negative linkage, any threat perceived as intended to compel you to do something that is good for you or that you feel you should do is likely to be suggestive of a positive linkage.

Although the law is crude, my impression is that it is reasonably accurate; phenotypes often indicate the underlying genotypes. Moreover, it is a synthesizing principle, which integrates and summarizes a wide range of social psychological phenomena. The typical effects of a given relationship tend to induce that relationship; similarly, it seems that any of the typical effects of a given relationship tend to induce the other typical effects. For example, among the typical effects of a cooperative relationship are positive attitudes, perception of similarities, open communication, and orientation toward mutual enhancement. One can integrate much of the literature on the determinants of positive and negative attitudes in terms of the other associated effects of cooperation and competition. Thus, positive attitudes result from perceptions of similarity, open communication, and so on. Similarly, many of the determinants of effective communication can be linked to the other typical effects of cooperation or competition, such as positive attitudes and power sharing.

3.6 Summary of the Theory of Conflict Resolution

In brief, the theory equates a constructive process of conflict resolution with an effective cooperative problem-solving process in which the conflict is the mutual problem to be resolved cooperatively. It also equates a destructive process of conflict resolution with a competitive process in which the conflicting parties are involved in a competition or struggle to determine who wins and who loses; often the outcome of the struggle is a loss for both parties. The theory further indicates that the typical effects of cooperation foster a cooperative-constructive process of conflict resolution. The theory of cooperation and competition outlined thus far in the chapter is a well-verified theory of the effects of cooperation and competition and thus allows insight into what can give rise to a constructive or destructive process.

The theory cannot serve as a cookbook for a practitioner in the field of conflict resolution. It is a general intellectual framework for understanding what goes on in conflicts and how to intervene in them. In addition, understanding and intervening in a specific conflict requires specific knowledge about the conflicting parties, their social and cultural contexts, their aspirations, their conflict orientations, the social norms, and so on.

Cooperation-competition, although of central importance, is only one factor influencing the course of conflict. The other chapters in this Handbook detail some of the other ingredients affecting conflict: power and influence, group problem solving, social perception and cognition, creativity, intrapsychic conflict, and personality. A practitioner must develop a mosaic of theories relevant to the specific situation of interest rather than relying on any single one. The symptoms or difficulties in one situation may require emphasis on the theoretical theme related to power; in another, it may require focusing on problem-solving deficiencies.

3.7 Implications of the Theory for Understanding Conflict

Kurt Lewin, a famous psychologist, used to tell his students, of whom I was one, that "there is nothing so practical as a good theory." To this point, I have presented the basic ideas of a good theory; in what follows, I indicate their usefulness in conflict situations.

3.7.1 The Importance of a Cooperative Orientation

The most important implication of cooperation-competition theory is that a cooperative or win-win orientation to resolving a conflict enormously facilitates constructive resolution, while a competitive or win-lose orientation hinders it. It is easier to develop and maintain a win-win attitude if you have social support for it. The social support can come from family, friends, coworkers, employers, the media, your community, or the culture in which you are embedded.

To have a win-win attitude in a hostile environment, it is valuable to become part of a network of people or a member of groups with similar orientations who can extend social support to you. It is also helpful to develop the personal strengths and skills that are useful in bucking the tide.

If you are the manager in a system (e.g., a principal in a school, a CEO in a company, a parent in a family), it is worthwhile to recognize that basic change in the system involves more than educating students, employees, or children to have a win-win orientation. It also involves educating yourself and other key people in the system, such as supervisors, staff, teachers, and parents, so that their actions reflect and support a win-win orientation. In addition, it often requires fundamental change in the incentive structure so that the rewards, salaries, grades, perks, and so on in the system do not foster a win-lose relationship among people.

3.7.2 Reframing

The second most important implication of the theory has to do with the cooperative process involved in constructive conflict resolution. At the heart of this process is reframing the conflict as a mutual problem to be resolved (or solved) through joint cooperative efforts. Reframing helps to develop a cooperative orientation to the conflict even if the goals of the conflicting parties are initially seen to be negatively interdependent. A cooperative orientation to what is initially a win-lose conflict leads the parties to search for just procedures to determine the winner, as well as for helping the loser gain through compensation or other means. Reframing assumes that whatever resolution is achieved, it is acceptable to each party and considered to be just by both. This assumption is made explicit when one or both parties to a conflict communicate to the other something like, “I won’t be satisfied with any agreement unless you also feel satisfied with it and consider it to be just, and I assume that you feel the same way. Is my assumption correct?”

Thus, consider a school that is developing site-based management (SBM) procedures but faces a conflict (the second opening vignette of the Introduction to this Handbook). One group of teachers, mainly white, insists on having teachers elected to the SBM executive committee from the various academic departments by majority vote. Another group of teachers, the Black Teachers Caucus (BTC), demands that several members of the committee be from minority groups to represent their interests. This conflict can be reformulated as a joint problem: how to develop SBM procedures that empower and are responsive to the interests and needs of faculty, parents, and students from minority groups without abandoning the regular democratic procedures whereby teachers are elected to the SBM committee by their respective departments.

This joint problem is not easy to solve, but many organizations have faced and resolved similar problems. There is reason to believe that if the conflicting groups—the SBM committee members elected by their departments and the BTC—define the conflict as a joint problem to be resolved cooperatively, they can come up with a solution that is mutually satisfactory. (See Chap. 2 for a discussion of resolving conflicts about what is just.)

3.7.3 The Norms of Cooperation

Of course, the parties are more likely to succeed in reframing their conflict into a mutual problem if the participants abide by the norms of cooperative behavior, even when in conflict, and have the skills that facilitate effective cooperation. The norms of cooperative behavior are similar to those for respectful, responsible, honest, empowering, and caring behavior toward friends or fellow group members. Some of these norms, particularly relevant to conflict, are the following:

- Placing the disagreements in perspective by identifying common ground and common interests.
- When there is disagreement, addressing the issues and refraining from making personal attacks.
- When there is disagreement, seeking to understand the other's views from his or her perspective; trying to feel what it would be like if you were on the other's side.
- Building on the ideas of the other, fully acknowledging their value.
- Emphasizing the positive in the other and the possibilities of constructive resolution of the conflict. Limiting and controlling expression of your negative feelings so that they are primarily directed at the other's violation of cooperative norms (if that occurs) or at the other's defeatism.
- Taking responsibility for the harmful consequences—unwitting as well as intended—of what you do and say; seeking to undo the harm as well as openly accepting responsibility and making sincere apology for it.
- If the other harms you, be willing to forgive if the other accepts responsibility for doing so, sincerely apologizes, and is willing to try to undo it; seeking reconciliation rather than nurturing an injury or grudge.
  Being responsive to the other's legitimate needs.
- Empowering the other to contribute effectively to the cooperative effort; soliciting the other's views, listening responsively, sharing information, and otherwise helping the other—when necessary—to be an active, effective participant in the cooperative problem-solving process.

Being appropriately honest. Being dishonest, attempting to mislead or deceive, is of course a violation of cooperative norms. However, one can be unnecessarily and inappropriately truthful. In most relationships, there is usually some ambivalence, a mixture of positive as well as negative thoughts and feelings about the other and about oneself. Unless the relationship has developed to a very high level of intimacy, communicating every suspicion, doubt, fear, and sense of weakness one has about oneself or the other is likely to be damaging to the relationship—particularly if the communication is blunt, unrationalized, and unmodulated. In effect, one should be open and honest in communication but appropriately so, realistically taking into account the consequences of what one says or does not say and the current state of the relationship.

- Throughout conflict, remaining a moral person—person who is caring and just—and considering the other as a member of one's moral community—therefore, as a person who is entitled to care and justice.

In the heat of conflict, there is often a tendency to violate the norms of cooperation. For example, you begin to attack the other as a person ("you're stubborn," "you're selfish," "you're unreasonable," "you're inconsiderate," "you're narcissistic," "you're paranoid"). Recognize when you start to do this, stop, apologize, and explain what made you angry enough to want to belittle and hurt the other. If the other starts to do this to you, then interrupt, explain why you are interrupting, and try to resume a mutually respectful dialogue: "You're calling me names; that's making me angry and makes me want to retaliate, so pretty soon we'll be in a name-calling contest and that will get us nowhere. Let's stick to the issues and be respectful of one another. If you're angry with me, tell me why. If I'm at fault, I'll remedy it."

It is wise to recognize that you, as well as the other, have hot buttons that, if pressed, are likely to evoke strong emotions. The emotions evoked can be anxiety, anger, rage, fear, depression, withdrawal, and so on. It is important to know your own hot buttons and how you tend to react when they are pressed, so that you can control your reactions in that event. Sometimes you need to take time out to control your emotional reactions and consider an appropriate response to what elicits them. Similarly, it is valuable to know the other's hot buttons so as to avoid pressing them and provoking disruptive emotions in the other.

3.7.4 The Values Underlying Constructive Conflict Resolution

The norms of cooperation and constructive conflict resolution reflect some basic values, to which people who are "profoundly divided by reasonable religious, philosophical, and moral doctrines" can adhere (Rawls 1996: xxxix). A reasonable doctrine includes conceptions of the values and norms with regard to conflict that people who adhere to another reasonable doctrine (as well as those who adhere to one's own) can endorse and be expected to follow during conflict. Thus, pro-life and pro-choice advocates in the abortion conflict may have profoundly differing views, but they are both components of reasonable doctrines if the adherents to each are willing to follow common values in dealing with their conflict about abortion. Among such values are reciprocity, human equality, shared community, fallibility, and nonviolence. A brief discussion of these interrelated values follows.

Reciprocity. This is the value in the maxim, "Do unto others as you would have others do unto you." My understanding of the maxim as it applies to conflict requires each party to treat the other with the fairness that it would normatively expect if in the other's position. It assumes reciprocity from the other—fairness to and from the other. The fairness in behavior, process, and outcomes expected is normative. As defined by one's culture, it is how the conflicting parties should or should not behave toward one another if they are, at a minimum, to avoid a destructive conflict or, more positively, to promote constructive management of their conflict. The norms against violence, disrespect, deceit, and irresponsibility are widespread standards for avoiding destructive conflict.

Human Equality. This value implies that all human beings are equally entitled to just and respectful treatment, with consideration for their needs and to such basic liberties as freedom of conscience, thought, and expression, as well as freedom from coercion. You are entitled to this from the other, and the other is entitled to this from you. Human equality does not imply that people necessarily have the same status, privileges, power, needs, or wealth. It does imply that such differences are not the consequence of one's violation of the other's entitlements.

Shared Community. Implicit in constructive conflict resolution is mutual recognition of being part of a broader community that members wish to preserve, a community sharing some key values and norms. Such recognition occurs despite important differences between oneself and the other.

Fallibility. The sources of disagreement between reasonable people are manifold. Disagreements may arise from such sources as the nature of the evidence, the weight to be given to types of evidence, and the vagueness of the moral or other concepts involved, as well as from differences in basic values or worldviews. Reasonable people understand that their own judgment as well as the judgment of others may be fallible.

Nonviolence. This value implies that neither you nor the other use coercive tactics to obtain agreement or consent. Such tactics include physical or psychological violence (e.g., humiliation), destruction of property or other valued goods, harm to one's life chances (a potential career), and harm to one's loved ones.

3.8 Implications for Managing Conflict

In prior sections, discussion focused on the attitudes, norms, and values that foster cooperation. These are necessary but not in themselves sufficient. Knowledge and skills are also important in promoting constructive resolution of a conflict. This is the thesis underlying this Handbook. Knowledge of the theory offers a useful framework for organizing one's thinking about the social psychological consequences of cooperation and competition, as well as the conditions that lead to one rather than the other. It is a way of orienting oneself to new situations. Along with the other theories discussed in this book, it enlarges one's knowledge of the range of conditions to be considered as one wishes to develop and maintain a constructive, cooperative process of conflict resolution and prevent developing a destructive process.

Skills are also vitally important for developing and implementing successfully an effective, cooperative problem-solving process. There has not been much systematic discussion of the skills involved in constructive solutions to conflict. There are, I believe, three main kinds useful to the participants in a conflict as well as to third parties (such as mediators, conciliators, counselors, or therapists) who are called on to provide assistance to conflicting parties. For convenience, I label them rapport-building skills, cooperative conflict resolution skills, and group process and decision-making skills.

First, there are the skills involved in establishing effective working relationships with each of the conflicting parties and between the conflicting parties if you are the mediator or with the other if you are a participant. Some of the components of this broad category include such skills as breaking the ice; reducing fears, tensions, and suspicion; overcoming resistance to negotiation; establishing a framework for civil discourse and interaction; and fostering realistic hope and optimism. Thus, before negotiations begin between two individuals or groups perceiving each other as adversaries, it is often useful to have informal social gatherings or meetings in which the adversaries can get to know one another as human beings who share some similar interests and values. Skill in breaking the ice and creating a safe, friendly atmosphere for interaction between the adversaries is helpful in developing the pre-negotiation experiences likely to lead to effective negotiations about the issues in dispute.

A second, related set of skills concerns developing and maintaining a cooperative conflict resolution process among the parties throughout their conflict. These are the skills that are usually emphasized in practicum courses or workshops on conflict resolution. They include identifying the type of conflict in which you are involved; reframing the issues so the conflict is perceived as a mutual problem to be resolved cooperatively; active listening and responsive communication; distinguishing between needs and positions; recognizing and acknowledging the other's needs as well as your own; encouraging, supporting, and enhancing the other; taking the perspective of the other; identifying shared interests and other similarities in values, experiences, and so on; being alert to cultural differences and the possibilities of misunderstanding arising from them; controlling anger; dealing with difficult conflicts and difficult people; being sensitive to the other's anxieties and hot buttons and how to avoid pressing them; and being aware of your own anxieties and hot buttons as well as your tendencies to be emotionally upset and misperceiving if they are pressed so that these can be controlled.

A third set of skills is involved in developing a creative and productive group problem-solving and decision-making process. These include skills pertinent to group process, leadership, and effective group discussion, such as goal and standard setting; monitoring progress toward group goals; eliciting, clarifying, coordinating, summarizing, and integrating the contributions of the various participants; and maintaining group cohesion. This third set also includes such problem-solving and decision-making skills as identifying and diagnosing the nature of the problem confronting the group; acquiring the relevant information necessary for developing possible solutions; creating or identifying several possible alternative solutions; choosing the criteria for evaluating the alternatives (such as the 'effects' on economic costs and benefits, on relations between the conflicting parties, and on third parties); selecting the alternative that optimizes the results on the chosen criteria; and implementing the decision through appropriate action.

People are not novices with regard to conflict. From their life experiences, many have developed some of the component skills involved in building rapport, constructive conflict resolution, and effective group process and problem solving. However, some are not aware that they have the skills or how and when to use them in a conflict. Everyone has been a participant and observer in many conflicts from childhood on. As a result, we posses implicit knowledge, preconceptions, attitudes, and modes of behavior toward conflict that may be deeply ingrained before any systematic training occurs. Many of a person's preexisting orientations to conflict and modes of behavior in it reflect those prevalent in his or her culture, but some reflect individual predispositions acquired from unique experiences in the contexts of family, school, watching TV, and the like.

Before students can acquire explicit competence in conflict resolution, they have to become aware of their preexisting orientations to conflict as well as their typical behaviors. Having a model of good performance develops awareness and motivation that students can compare with their preconscious, preexisting one. Internalization comes from guided and repeated practice in imitating the model. Feedback on the students' success gradually shapes their behavior to be consistent with the model, and frequent practice leads to its internalization. Once the model has been internalized, recurrence of earlier incompetent orientations to conflict is experienced as awkward and out of place because there are internal cues to the deviations of one's behavior from the internalized model. In tennis, if you have internalized a good model of serving, internal cues tell you if you are deviating from it (say, by throwing the ball too high). If self-taught tennis students have internalized poor serving models, training should be directed at making them aware of this and providing a good model. So too in conflict resolution.

In summary, the discussion in this and the preceding sections has centered on the orientation, norms, values, and skills that help to develop a cooperative, constructive process of conflict resolution. Without competence in the skills, having a cooperative orientation and knowledge of conflict processes is often insufficient to develop a cooperative process of conflict resolution. Similarly, having the skills is insufficient to develop a cooperative process without the cooperative orientation and motivation to apply the skills or without the knowledge of how to apply the skills in various social and cultural contexts.

3.9 Implications for Training

The material already presented in this chapter has several implications for training. They center on the social context of learning, the social context of applying one's learning, the substantive content of the training, and the reflective practitioner.

3.9.1 The Social Context of Learning

The theory described in this chapter suggests that the social context of learning be one in which cooperation, constructive conflict resolution, and creative controversy are strongly emphasized. The teaching method employed should take the form of cooperative learning, and the conflictual interactions within the classroom or workshop between teacher and students and among students should model those of creative controversy and constructive conflict resolution. The social context of learning should walk the talk, and in so doing offer students the experiences that support a cooperative orientation, exemplify the values and social norms of cooperation, and model the skills in constructive management of conflict.

3.9.2 The Social Context of Application

It can be anticipated that many social contexts are unfavorable to a cooperative orientation and the use of one's skills in constructive conflict resolution. In some social contexts, an individual who has such skills may expect to be belittled by friends or associates as being weak, unassertive, or afraid. In other contexts, she may anticipate accusations of being 'disloyal,' a 'traitor,' or an "enemy lover" if she tries to develop a cooperative problem-solving relationship with the other side. In still other contexts, the possibility of developing a constructive conflict resolution process seems so slim that one does not even try to do so. In other words, if the social context leads you to expect to be unsuccessful or devalued in employing your skills, you are not likely to use them; you will do so if it leads you to expect approval and success.

This explanation suggests that in unfavorable social contexts, skilled conflict resolvers often need social support as well as two additional types of skill. One relates to the ability to place yourself outside or above your social context so that you can observe the influences emanating from it and then consciously decide whether to resist them personally. The other type involves the skills of a successful change agent—someone who is able to help an institution or group change its culture so that it facilitates rather than hinders constructive conflict resolution. I mention these additional skills because it is important to recognize that institutional and cultural changes are often necessary for an individual to feel free to express his or her constructive potential.

The common need for social support after training has implications for who are selected for training and also for post-training contacts. There are several ways to foster a social context that is supportive: train all of the participants, train the influential people, or train a cohort of people of sufficient size to provide effective mutual support in the face of resistance. Post-training contacts with the training institution and its trainers may also yield the social support necessary to buttress the individual in a hostile environment.

3.9.3 The Substantive Content of Training

In prior sections of this chapter, I have outlined what I consider to be the attitudes, knowledge, and skills that amount to a framework for education in constructive conflict resolution. A skillful trainer fleshes out such a framework with substantive content that is sufficiently vital and intellectually compelling to engage the interest and motivation of the student, is relevant to his or her most common and most difficult conflicts, and is sufficiently diverse in content and social context to facilitate generalizing and applying the training in a variety of situations. To accomplish these objectives, a trainer must not only have a clear framework for training, but must also be open and creative so that he or she can respond to the students' needs effectively.

3.9.4 The Reflective Practitioner

One of the important goals of education in this area is to help the student, as well as the trainer become a reflective practitioner of constructive conflict resolution. I refer to two kinds of reflection: on managing the conflicts you are experiencing and on the framework of conflict resolution you are employing. Self-reflection about how you are handling conflicts is necessary to continuing improvement and also to prevent old habits, your hot spots, social pressure, and the like from making you regress to less constructive modes of conflict resolution.

Conflict resolution as a field of study is relatively young; it is going through a period of rapid intellectual development. It is experiencing an upsurge in research, theoretical development, and practical experience that I hope results in improvement of the frameworks used for training in conflict resolution. The reflective practitioner, by reflecting on his or her practice, can learn from as well as contribute to this growing body of knowledge and reflected-on experience.

3.9.5 Suppose the Other Does not Want to Cooperate: What Then?

Suppose the other wants to win and does not want to cooperate to resolve the conflict constructively. What then? Or suppose the other agrees to negotiate a resolution of the conflict but engages in dirty tricks to try to triumph in the negotiations. How do you respond? These are difficult questions, and it should be clear that in some instances, it might be impossible to establish a cooperative conflict resolution process or prevent the other from employing dirty tricks during a negotiation. Nevertheless, as the cases in the Introduction to this Handbook indicate, difficult, deep-rooted conflicts can be resolved or managed well. I next briefly discuss some suggestions for managing each of the two difficult types of situations.

3.9.5.1 The Other Refuses to Cooperate

There are two main reasons for not wanting to cooperate: (1) you think it would be futile, a waste of time and energy, or (2) you feel you are the dominant power and are satisfied with the existing situation and will lose something of value (e.g., power, status, identity, wealth, religious doctrine) if you do. Before attempting to influence the other in either case, it is crucial to seek to understand the other—the other's position, reasons, emotions, social context, and experiences that have led to and support the other's position. This requires the development of communication with the other and active, nonjudgmental listening to the other. After achieving some understanding of the other, one will seek to influence the other to be willing to cooperate; influence attempts commonly involve the use of persuasive strategies or nonviolent power strategies, or both.

Persuasive strategies involve three types of appeals: to moral values, self-interest, or self-fulfillment.

A *moral appeal* to another person (group, organization, or nation) who feels it is futile to attempt cooperation might be: "If you are a moral person, you should try to achieve a good outcome even if it is difficult or the chances of success are small. If you see a child drowning near you, you should try to rescue him even if the chances of success are small and it is difficult to do. Similarly, it is your moral obligation to try to resolve your conflict with the other in a cooperative manner even though you think the chances of success are small and it may be a difficult process."

Appeal to the moral values of the dominant power assumes he or she is not fully aware of the negative impact of their power on the low-power person or group. For example, one might appeal to values related to justice, religion, or the welfare of one's grandchildren, to name a few. Engaging high-power members to see the discrepancy between their practices and their moral values or conscience could move them to take action and change their behavior.

Self-interest appeals emphasize the gains that can be obtained and losses that can be prevented when there is cooperation to resolve the conflict. It is important that such messages be carefully constructed so as to clearly state the specific actions and changes requested of the other and to highlight the values and benefits to the other by cooperating and the potential losses of not cooperating (Deutsch 2006).

Appeals to *self-actualization* focus on enhancing the sense that one's better self—a self that one has wanted to be—is being actualized. In a sense, these are a type of self-interest appeal. The gain for the other is the feelings associated with an actualized self. In considering ways that one might share one's power over others, one might emphasize the use of one's power to further common interests; the spiritual emptiness of power *over* others; the fulfillment of creating something that goes well beyond self-benefit. By creating power *with* others rather than maintaining noncooperation or power *over* (Follett 1973), you may actually increase your power.

Low-power individuals or groups seeking change in those who have a vested interest in maintaining their power sometimes find it difficult to employ persuasion strategies because of rage or fear. Rage, as a result of the injustices they have experienced, may lead them to seek revenge, to harm or destroy those in power. Fear of the powerful to inflict bearable harm may inhibit efforts to bring about change in the powerful.

Given the possibility of the prevalence of rage or fear among low-power groups, it would be the goal of change agents (group leaders, mediators, conciliators, therapists) who seek to foster cooperation, rather than rage or fear, to harness the energy created by feelings of rage and fear and convert it into effective cooperative action. (See Gaucher and Jost 2011.) By engaging large numbers of people through social media and other communication methods, you channel the energy generated by feelings of rage or fear toward effective action. Here the task of the change agent is to help people realize that they are more likely to achieve their goals through effective action, including cooperation with potential allies among members of high-power groups. It is important for the change agent to recognize the power of the motivational energy of low-power groups, regardless of its source.

A potentially effective strategic starting point using persuasive strategies would be for low-power groups to use social influence strategies by seeking out and creating alliances with those members of high-power groups, as well as other prestigious and influential people and groups, who are sympathetic to their efforts of building cooperation (Deutsch 2006). Developing allies is a key method of increasing a low-power group's power and increasing its influence and credibility with those in power.

It is useful for change agents to understand the psychological implications of appealing to the power needs of members of high-power groups—understanding how to convince those in power that their power needs can be fulfilled through fostering a common good.

Nonviolent power strategies involve enhancing one's own power (by developing the latent power in one's self and one's group, as well as developing allies), employing the power of the powerful against the powerful, and reducing the power of the powerful. Sharp (1973, 2005) has elaborated in great detail the many tactics available to those who seek to employ nonviolent power strategies and also discussed the strategy in producing successful nonviolent change opposing dominating, exploiting others. There are three types of nonviolent actions:

1. *Acts of protest* such as the recent events in the Middle East.
2. *Noncooperation* such as in Aristophanes' *Lysistrata* when the women withhold sex from their spouses until war is abolished.
3. *Nonviolent intervention* such as general strikes and other methods of disrupting the economy and other components of the status quo.

The employment of nonviolent methods against a potentially violent, autocratic, resistance to change in power often requires considerable courage, discipline, and stamina as well as effective preplanning and organization.

There is a difference between persuasive strategies and nonviolent strategies. Nonviolent strategies are often used when persuasion strategies by themselves are not effective in bringing about change. The aim of nonviolent strategies is to open those in power so that they can be persuaded to change: resistance to and interference with the implementation of the power of the high-power group makes its power ineffective and could open it to the possibility of persuasion. Both are useful in altering the status quo in service of developing cooperation. However, in contrast to violent strategies, neither persuasion nor nonviolence seeks to destroy those in high power: they seek to change the relationship so that power is shared and used to benefit both sides.

There are two major problems with the use of violence: it commonly leads to increasing destructive cycles of reciprocating violence between the conflicting parties, and it can transform those using violent methods into mirror images of one another, so if a low-power group employs violence to overthrow a tyrannical high-power group, it may become tyrannical itself. I am suggesting that violence is never necessary to stop unrelenting tyranny. As Mandela (1995) indicated, if violence is thought to be necessary to motivate the other, it should be employed only against nonhuman targets, such as bridges or communication facilities, only.

3.9.5.2 Facing Dirty Tricks During Cooperative Negotiation

Suppose the other agrees to negotiate cooperatively to resolve the conflict but engages in dirty tricks to advantage itself during negotiations, such as lying, misrepresenting, spreading false rumors, undermining your power, or amassing its own power to threaten and coerce you. What do you do? First, you openly confront the other with what you consider to be his dirty trick in a no antagonistic manner and give the other a chance to respond and explain. He might persuade you that you are mistaken, and if so, you would apologize. If he denies guilt but you are not convinced of his innocence, you seek to resolve this conflict cooperatively. Here the involvement of neutral third parties such as a judge, mediator, or therapist may be of value or necessary. If the other pleads guilty, apologizes, and pledges not to continue to engage in dirty tricks but you are not completely reassured, it may be necessary to establish a mutually agreed- on neutral, independent individual or system that can detect dirty tricks (by you or the other) as well as verify or falsify accusations of dirty tricks and provide sufficient positive and negative incentives to deter their occurrence.

Whether or not the other is willing to engage in fair cooperation, one's own approach throughout should employ the four Fs: be firm, fair, flexible, and friendly;

- *Firm* in the sense that you will strongly protect yourself from being disadvantaged unfairly.
- *Fair*, in the sense that you will treat the other fairly and not attempt to disadvantage the other by dirty tricks.
- *Flexible* in the sense that you will not commit yourself to rigid positions and will respond flexibly to the legitimate interests of the other.
- *Friendly*, in the sense that you are always open, even after some difficulties, to fair, amiable, mutual cooperation.

3.10 Conclusion

The central theme of this chapter is that a knowledgeable, skillful, cooperative approach to conflict enormously facilitates its constructive resolution. However, there is a two-way relation between effective cooperation and constructive conflict resolution. Good cooperative relations facilitate constructive management of conflict. The ability to handle constructively the inevitable conflicts that occur during cooperation facilitates the survival and deepening of cooperative relations.

References

Deutsch, M., 1949a: "An Experimental Study of the Effects of Cooperation and Competition upon Group Processes", in: *Human Relations*, 2: 199–231.

Deutsch, M., 1949b: "A Theory of Cooperation and Competition", in: *Human Relations*, 2: 129–151.

Deutsch, M., 1973: *The Resolution of Conflict: Constructive and Destructive Processes* (New Haven, CT: Yale University Press).

Deutsch, M., 1985: *Distributive Justice: A Social Psychological Perspective* (New Haven, CT: Yale University Press).

Deutsch, M., 2006: "A Framework for Thinking About Oppression and Its Change", in: *Social Justice Research*, 19,1: 7–41.

Deutsch, M., 2011: "Cooperation and Competition", in: Coleman, Peter T. (Ed.): *Conflict, Interdependence, and Justice: The Intellectual Legacy of Morton Deutsch* (New York: Springer).

Follett, M.P., 1973: "Power", in: Fox E.M.; Urwick, L. (Eds.): *Dynamic Administration: The Collected Papers of Mary Parker Follett* (London: Pitman) (Originally published 1924).

Gaucher, D.; Jost, J.T., 2011: "Difficulties Awakening the Sense of Injustice and Overcoming Oppression: On the Soporific Effects of System Justification", in: Coleman, Peter T. (Ed.):

Conflict, Interdependence, and Justice: The Intellectual Legacy of Morton Deutsch (New York: Springer).

Johnson, D.W.; Johnson, R.T., 2005: "New Developments in Social Interdependence Theory", in: *Psychology Monographs*, 131: 285–358.

Johnson, D.W.; Johnson, R.T., 2011: "Intellectual Legacy: Cooperation and Competition", in: Coleman, Peter T. (Ed.): *Conflict, Interdependence, and Justice: The Intellectual Legacy of Morton Deutsch* (New York: Springer).

Mandela, N., 1995: *A Long Walk to Freedom: The Autobiography of Nelson Mandela* (London: Little, Brown).

Rawls, J., 1996: *Political Liberalism* (New York: Columbia University Press).

Sharp, G., 1973: *The Politics of Nonviolent Action* (Boston: Porter Sargent Publishers).

Sharp, G., 2005: *Waging Nonviolent Struggle: Twentieth Century Practice and Twenty-First Century Potential* (Boston: Porter Sargent Publishers).

Chapter 4
A Framework for Thinking About Oppression and Conflict

4.1 Introduction

This paper[1] provides a framework for thinking about oppression and how to overcome it.[2] It considers the value premise underlying the use of the term 'oppression.' It then discusses the nature of oppression, the forms it takes, and what keeps it in place. In its final two sections, it focuses on awakening the sense of injustice and the strategies and tactics for overcoming injustice.

In this paper, my purpose is to provide a framework for thinking about oppression and how to overcome it. Oppression is, I believe, at the root of many of the most serious, enduring conflicts in the world today.

The paper is divided into the following sections. The first considers the value premise underlying my use of the term 'oppression' (4.2). The second is a

---

[1] This text was first published as: Deutsch, M. (2006). A Framework for Thinking about Oppression and Its Change. *Social Justice Research,* 19(1), 7–41. Print ISSN: 0885-7466, Online ISSN: 1573-6725 Permission to republish this text was granted by Springer Permissions in Heidelberg on 11 November 2014.

[2] An earlier version of this paper, entitled "Oppression and Conflict," was presented as a plenary address at the Annual meeting of the International Society of Justice Research in Skovde, Sweden on June 17, 2002. This paper was the starting point of an ongoing, informal seminar on social justice held at Teachers College, Columbia University. The other participants in the seminar included Peter Coleman, Michelle Fine, Beth Fisher-Yoshida, Janet Gerson, Eric Marcus, Susan Opotow, Ellen Raider, Esther Salomon, Janice Steil, and Melissa Sweeney. From the discussions of theory, research, and practice during the meetings of the seminar, emerged plans for a Conference on Interrupting Oppression and Sustaining Justice. Its aim was to stimulate interaction about overcoming oppression among scholars from different academic fields, social and political activists, and graduate students in different disciplines. The Conference took place on February 27 and 28, 2004 at Teachers College. This issue of *Social Justice Research* presents some papers that were prepared before the Conference (and revised afterwards) and some that emerged from the Conference. Other papers from the Conference can be found under the heading of IO & SJ at the website of the International Center for Cooperation and Conflict Resolution of Teachers College: <http://www.tc.edu/icccr/>.

P.T. Coleman and M. Deutsch, *Morton Deutsch: A Pioneer in Developing Peace Psychology*, SpringerBriefs on Pioneers in Science and Practice 30,
DOI 10.1007/978-3-319-15440-4_4

discussion of the nature of oppression (4.3). The third addresses the question, "What forms does oppression take?" (4.4). The fourth asks, "What keeps oppression in place?" (4.5). The fifth addresses the awakening of the sense of injustice (4.6). The sixth provides a discussion of the strategies and tactics for overcoming oppression, which often involve violent conflict with groups in power (4.7). In the final section of my paper, I will discuss some nonviolent strategies and tactics for overcoming oppression (4.8).

My discussion will not focus on the different contexts in which oppression occurs, such as the family, work, education, and between ethnic, religious, and racial groups. There is an excellent discussion of the different contexts of oppression in the book, *Social Inequality* (Neckerman 2004), which presents extensive empirical data about inequality in various contexts.

4.2 The Value Premise Underlying My Use of the Term Oppression

The use of such terms as 'oppression' and 'injustice' implies the existence of a violation of a value or set of values. In employing such terms, I am adhering to the values incorporated in the Universal Declaration of Human Rights adopted by the United Nations on December 10, 1948. I believe there are several central values underlying the Universal Declaration: *democracy*, *egalitarianism*, and *effective cooperation*. Democratic egalitarianism pervades the thirty articles of the Declaration. Effective cooperation among and within nations (i.e., among the individuals, groups, and institutions which comprise national and international groupings) is necessary to create the social, material, and environmental conditions that are conducive to democratic egalitarianism. Egalitarianism is expressed in the first sentence of the Preamble to the Declaration: "Whereas recognition of the inherent dignity and of the equal and inalienable rights of all members of the human family is the foundation of freedom, justice, and peace in the world…"

In my book, *Distributive Justice* (Deutsch 1985: 41–42), I have written about equality and egalitarianism as follows:

> The concept of equality has been discussed extensively by moral and political philosophers (Berlin 1955–1956; Tawney 1964, McCloskey 1966; Wilson 1966; Benn 1967; Pennock/Chapman 1967; Oppenheim 1968; Rawls 1971; Dworkin 1981a, b; Walzer 1983). In the vast literature dealing with equality, it is defined in various ways. I shall not attempt to summarize or critically examine this literature. My sense of it is that advocates of equality and egalitarianism are primarily opposed to invidious distinctions among people but do not assume that all distinctions are invidious. Invidious distinctions are ones that promote (1) generalized or irrelevant feelings of superiority-inferiority (if I am a better tennis player or more good-looking than you, I am superior to you as a person); (2) generalized or irrelevant status differences (if I am a manager and you are a worker in a factory, I should have a higher standard of living than you); (3) generalized or irrelevant superordinate-subordinate relations (if I am a captain and you are a private, I can order you to shine my shoes); or (4) the view that the legitimate needs and interests of some people are not as important or do

not warrant as much consideration as those of other people (this may be because of my sex, race, age, national or family origin, religion, political affiliation, occupation, or physical handicap, or because of special talents or lack of talent).

From an egalitarian perspective, one is not making an invidious distinction and thus creating inequality when one recognizes, approves, applauds, honors, or shows appreciation of an unusually good performance, of a courageous action, or of a well-accomplished difficult task. Equality does not imply identical treatment of everyone without regard to particular circumstance. Honoring another's performance does not diminish those who are not honored unless they consider themselves to be in a competitive or zero-sum relationship; if they view themselves as part of a cooperative community, they are enhanced by another's honor. The insistence on treating people identically, without regard to circumstance, is a pseudo egalitarianism, which often masks basic doubts or ambivalence about one's commitment to egalitarian values.

Egalitarianism is conducive to effective cooperation in that it promotes social harmony, which in turn promotes mutual aid and the efficient specialization of function in cooperative work. (See Deutsch (1990) for a discussion of the psychological consequences of different forms of social organization.)

Although cooperative democratic egalitarianism or the Universal Declaration of Human Rights does not imply that all individuals, groups, or categories of people are treated identically or have the same positions in society, it does imply that the material and social conditions that affect individual well-being are distributed so that there are not gross systematic disparities in well-being, opportunities for human development, or the rights of people, individually or collectively. Of course, due to inadequate knowledge, there is considerable uncertainty and conflict about how to create a world that would actualize the values of the Universal Declaration.

In this paper, my discussion of forms of oppression assumes that systematic, large inequalities of the kinds described earlier are unjust. Some would challenge this assumption and assert that such inequalities are inevitable and inherent in human nature as well as in the social orders of animals and humans, and that they also promote economic productivity. However, I have pointed out (Deutsch 1985: 40–41):

> If a cooperative system is oriented toward increasing its economic productivity, its rational tendency will be to allocate its economic functions and goods to use them effectively, but to allocate its rewards (consumer goods) according to need or equality (if more than a bare necessity is available). However, it is also suggested that inherent pathologies in the extension of economic values throughout a society or in the temptation to accumulate personal power may give rise to an equity principle that allocates economic rewards and political power as well as economic functions and goods to those who appear to contribute the most to the group. This equity principle, over the long run, is likely to be dysfunctional for groups, economically as well as socially. Economically, by allocating rewards and power disproportionately, it enables those who are in power to bias the system of allocation to perpetuate their disproportionate rewards and power even when they are no longer making relatively large contributions to the group's well-being. Socially, it tends to foster the introduction of economic values in all aspects of social life with a resultant loss to the quality of life. As Diesing has written, 'A person becomes alienated from his possessions and creations when he learns to regard them as utilities which have value because other people desire them; he becomes alienated from other people when they are perceived as competing with him for scarce goods; and he becomes alienated from himself when he sees his own values as a utility based on the desires of others' (1962: 93).

The fact that cooperative egalitarianism is found in many hunting-fishing-gathering societies as well as more complex societies indicates that oppressive inequalities are not a human inevitability (see Mead 1937; Gil 1998; Kemp/Fry 2004). Henry Levin's paper indicates that large, cooperative, egalitarian enterprises are both more productive and more humane than their traditional counterparts, which reject cooperative ownership and management. I found similar results in my review of egalitarian systems in the laboratory as well as in the social world in my book, *Distributive Justice* (Deutsch 1985). I also note that anthropological reviews of peaceful societies indicate that such societies are mainly cooperative and egalitarian (see Howell/Willis 1989; Kemp/Fry 2004; Encyclopedia of Selected Peaceful Societies, 2005, at <http://www.peacefulsocieties.org>).

4.3 What Is Oppression?

Oppression is the experience of repeated, widespread, systemic injustice. It need not be extreme and involve the legal system (as in slavery, apartheid, or the lack of a right to vote) nor violent (as in tyrannical societies). Harvey (1999) has used the term "civilized oppression" to characterize the everyday processes of oppression in normal life. Civilized oppression "is embedded in unquestioned norms, habits, and symbols, in the assumptions underlying institutions and rules, and the collective consequences of following those rules. It refers to the vast and deep injustices some groups suffer as a consequence of often unconscious assumptions and reactions of well-meaning people in ordinary interactions that are supported by the media and cultural stereotypes as well as by the structural features of bureaucratic hierarchies and market mechanisms" (Young 1990: 41).

We cannot eliminate this structural oppression by getting rid of the rulers or by making some new laws, because oppressions are systematically reproduced in the major economic, political, and cultural institutions. While specific privileged groups are the beneficiaries of the oppression of other groups, and thus have an interest in the continuation of the status quo, they do not typically understand themselves to be agents of oppression.

4.4 What Forms Does Oppression Take?

I consider here five types of injustices that are involved in oppression: distributive injustice, procedural injustice, retributive injustice, moral exclusion, and cultural imperialism. To identify which groups of people are oppressed and what forms their oppression takes, each of these five types of injustice should be examined. For a comprehensive discussion of social psychological research related to the topics discussed later, see Tyler et al. (1997).

4.4.1 Distributive Injustice

Under this section, I shall briefly consider the distribution of four types of capital (Perrucci/Wysong 1999): consumption, investment, skill, and social capital.

Consumption capital is usually thought of as "standard of living." It includes income as well as job and financial security and the amounts and types of food and water, housing, clothing, physical security, health care, education, sanitation, physical mobility (such as travel), recreation, and services that are available to members of a group. Clearly, there are gross differences in income and standards of living among the different nations, among the different ethnic groups within nations, among the different classes, and between the sexes. For example, compare Sudan with Canada, African-Americans with Euro-Americans, employees of General Motors and its executives, and females and males.

Sen, for example (in Sen/Dreze 1999: Chap. 7, 140, in the book titled *India, Economic Development, and Social Opportunity*) writes: "Women tend in general to fare quite badly in relative terms compared to men, even within the same families. This is reflected not only in such matters as education and opportunity to develop talents, but also in the more elementary fields of nutrition, health, and survival." He estimated that there are "more than a hundred million missing women," in Asia and North Africa, as a result of the unequal deprivations they suffer compared to men. In other words, the survival rates of women compared to men is considerably lower than could be expected when these are compared to the relative survival rates of men and women in Europe, North America, and sub-Saharan Africa, where the differences in consumption capital available to males and females are not as unequal.

Investment capital "is what people use to create more capital" (Perrucci/Wysong 1999: 10). Income is related to consumption capital and also to wealth, which, in turn, is related to investment capital. Generally, wealth is distributed more unequally than income. The inequalities among nations, within nations, among ethnic groups, among the social classes, between the physically impaired and unimpaired, and between the sexes are apt to be considerably greater with regard to investment than consumption capital. In 1998, in the United States, the top 10 % of the population possessed 68.7 % of the financial assets, while the bottom 90 % had only 31.3 % (Scholz/Levine 2004), and this discrepancy has undoubtedly increased since then.

Skill capital is the specialized knowledge, social and work skills, as well as the various forms of intelligence and credentials, that are developed as a result of education, training, and experiences in one's family, community, and work settings. As Perrucci/Wysong (1999: 14) point out: "The most important source of skill capital in today's society is located in the elite universities that provide the credentials for the privileged class. For example, the path into corporate law with six-figure salaries and million-dollar partnerships is provided by about two-dozen elite law schools where children of the privileged class enroll. Similar patterns exist for medical school graduates, research scientists, and those holding professional degrees in management and business. People in high-income and wealth-producing professions will seek to protect the market value not only for themselves, but also for their children, who will enter similar fields." It is evident that those in non-privileged groups in many societies will have much less opportunity to enter elite universities and to acquire the skills and credentials which would have high market value.

Social capital is the network of social ties (family, friends, neighbors, social clubs, classmates, acquaintances, etc.), which can provide information and access to jobs and to the means of acquiring the other forms of capital, as well as emotional and financial support. It is the linkage that one has or does not have to organizational power, prestige, and opportunities. The social capital that one can acquire and maintain is affected by such factors as one's family, social class, membership in particular ethnic and religious groups, age, sex, physical disability, and sexual orientation. In most societies, the ability to acquire and maintain social capital by those who are underclass or working class, disabled, elderly, members of minority, ethnic, religious or racial groups, or women is considerably more limited than the ability of dominant groups. Personality, undoubtedly, also plays a role: one could expect that individuals who are ambitious, sociable, intelligent, and personally attractive will acquire more social capital than will those who are not.

To sum up this section on distributive justice (Deutsch/Coleman 2000: 56): "Every type of system—from a society to a family—distributes benefits, costs, and harms (its reward systems are a reflection of this). One can examine the different forms of capital (consumption, investment, skill, social) and such benefits as income, education, health care, police protection, housing, and water supplies, and such harms as accidents, rapes, physical attacks, imprisonment, death, and rat bites, and see how they are distributed among categories of people: rich versus poor, males versus females, employers versus employees, Whites versus Blacks, heterosexuals versus homosexuals, police officers versus teachers, adults versus children. Such examination reveals gross disparities in distribution of one or another benefit or harm received by the categories of people involved. Thus, Blacks generally received fewer benefits and more harm than Whites in the United States. In most parts of the world, female children are less likely than male children to receive as much education or inherit parental property, and they are more likely to suffer from sexual abuse."

4.4.2 Procedural Injustice

In addition to assessing the fairness of the distribution of outcomes, individuals judge the fairness of the procedures that determine the outcomes (see Lind/Tyler 1988, for a comprehensive discussion of procedural justice). Research evidence indicates that fair treatment and procedures are a more pervasive concern to most people than fair outcomes. Fair procedures are psychologically important because they encourage the assumption that they give rise to fair outcomes in the present and will also in the future. In some situations, where it is not clear what "fair outcomes" should be, fair procedures are the best guarantee that the decision about outcomes is made fairly. Research indicates that one is less apt to feel committed to authorities, organizations, social policies, and governmental rules and regulations if the procedures associated within them are considered unfair. Also, people feel affirmed if the procedures to which they are subjected treat them with the respect and dignity they feel is their due; if so treated, it is easier for them to accept a disappointing outcome.

Questions with regard to the justice of procedures can arise in various ways. Let us consider, for example, evaluation of teacher performance in a school. Some questions immediately come to mind: Who has 'voice' or representation in determining whether such evaluation is necessary? How are the evaluations to be conducted? Who conducts them? What is to be evaluated? What kind of information is collected? How is its accuracy and validity ascertained? How are its consistency and reliability determined? What methods of preventing incompetence or bias in collecting and processing information are employed? Who constitutes the groups that organize the evaluations, draw conclusions, make recommendations, and make decisions? What roles do teachers, administrators, parents, students, and outside experts have in the procedures? How are the ethicality, considerateness, and dignity of the process protected?

Implicit in these questions are some values with regard to procedural justice. One wants procedures that generate relevant, unbiased, accurate, consistent, reliable, competent, and valid information and decisions as well as polite, dignified, and respectful behavior in carrying out the procedures. Also, voice and representation in the processes and decisions related to the evaluation are considered desirable by those directly affected by the decision. In effect, fair procedures yield good information for use in the decision-making processes, voice in the processes for those affected by them, and considerate treatment as the procedures are being implemented (Deutsch/Coleman 2000: 44–45).

One can probe a system to determine whether it offers fair procedures to all. Are all categories of people treated with politeness, dignity, and respect by judges, police, teachers, administrators, employers, bankers, politicians and others in authority? Are some but not others allowed to have a voice and representation, as well as adequate information, in the processes and decisions that affect them?

It is evident that those people and groups with more capital are more likely to have access to political leaders and to be treated with more respect by the police, judges, and other authority figures than those with less capital. Also, their ability to have 'voice' in matters that affect them is considerably greater.

4.4.3 Retributive Injustice

Retributive injustice is concerned with the behavior and attitudes of people, especially those in authority, in response to moral rule breaking. One may ask: Are 'crimes' by different categories of people less likely to be viewed as crimes, to result in an arrest, to be brought to trial, to result in conviction, to lead to punishment or imprisonment or the death penalty, and so on? Considerable disparity is apparent between how "robber barons" and ordinary robbers are treated by the criminal-justice system, between manufacturers who knowingly sell injurious products (obvious instances being tobacco and defective automobiles) and those who negligently cause an accident. Similarly, almost every comparison of the treatment of Black and White criminal offenders indicates that, if there is a difference, Blacks receive worse treatment.

4.4.4 Moral Exclusion

Moral exclusion is about who is and is not entitled to fair outcomes and fair treatment by inclusion or lack of inclusion in one's moral community. Albert Schweitzer included all living creatures in his moral community, and some Buddhists include all of nature. Most of us define a more limited moral community.

Individuals and groups who are outside the boundary in which considerations of fairness apply may be treated in ways that would be considered immoral if people within the boundary were so treated. Consider the situation in Bosnia. Prior to the breakup of Yugoslavia, the Serbs, Muslims, and Croats in Bosnia were more or less part of one moral community and treated one another with some degree of civility. After the start of civil strife (initiated by power-hungry political leaders), vilification of other ethnic groups became a political tool, and it led to excluding others from one's moral community. As a consequence, the various ethnic groups committed the most barbaric atrocities against one another. The same thing happened with the Hutus and Tutsis in Rwanda and Burundi.

At various periods in history and in different societies, groups and individuals have been treated inhumanly by other humans: slaves by their masters, natives by colonialists, Blacks by Whites, Jews by Nazis, women by men, children by adults, the physically disabled by those who are not, homosexuals by heterosexuals, political dissidents by political authorities, and one ethnic or religious group by another.

When a system is under stress, are there differences in how categories of people are treated? Are some people more apt to lose their jobs, be excluded from obtaining scarce resources, or be scapegoated and victimized? During periods of economic depression, social upheaval, civil strife, and war, frustrations are often channeled to exclude some groups from the treatment normatively expected from others in the same moral community.

Moral exclusion "is perhaps the most dangerous form of oppression" (Young 1990: 53). It has led to genocide against the Jews and gypsies by the Nazis, the Turkish genocide of the Armenians, the autogenocide by the Khmer Rouge in Cambodia, the mass killings of the political opposition by the Argentinian generals, widespread terrorism against civilians by various terrorist groups, and the enslavement of many Africans, to mention only a few examples of the consequences of moral exclusion.

Lesser forms of moral exclusions and marginalization occur also against whole categories of people—women, the physically impaired, the elderly, and various ethnic, religious, and racial groups—in many societies where barriers prevent them from full participation in the political, economic, and social life of their societies. The results of these barriers are not only material deprivation but also disrespectful, demeaning, and arbitrary treatment as well as decreased opportunity to develop and employ their individual talents. For extensive research and writing in this area, see the work of Opotow (1987, 1990, 1995, 1996a, b, 2001).

4.4.5 Cultural Imperialism

"*Cultural Imperialism* involves the universalization of a dominant group's experience and culture and establishing it as the norm" (Young 1990: 59). Those living under cultural imperialism find themselves defined by the dominant others. As Young (*op. cit*) points out: "Consequently, the differences of women from men, American Indians or Africans from Europeans, Jews from Christians, becomes reconstructed as deviance and inferiority." To the extent that women, Africans, Jews, Muslims, homosexuals, etc. must interact with the dominant group whose culture mainly provides stereotyped images of them, they are often under pressure to conform to and internalize the dominant group's images of their group.

Culturally dominated groups often experience themselves as having a double identity, one defined by the dominant group and the other coming from membership in one's own group. Thus, in my childhood, adult African-American men were often called 'boy' by members of the dominant White groups but within their own group, they might be respected ministers and wage earners. Culturally subordinated groups are often able to maintain their own culture because they are segregated from the dominant group and have many interactions within their own group, which are invisible to the dominant group. In such contexts, the subordinated culture commonly reacts to the dominant culture with mockery and hostility fueled by their sense of injustice and of victimization.

4.5 What Keeps Oppression in Place?

Here I consider other factors that that contribute to the maintenance of oppression: the superior power of the dominant group; the social production of meaning in the service of legitimating oppression; the self-fulfilling prophecies arising from oppression; and the distorted relationship between the oppressed and the oppressor.

4.5.1 The Superior Power of the Oppressor

Elsewhere, I have discussed different forms of power (Deutsch 1973), as have many others. Here, I am focusing on 'competitive' power, the power to control, dominate, or exploit another person, group, or nation whose power is not sufficient to prevent such domination or exploitation, rather than on 'cooperative' power, where it is to the benefit of each other if the other's power is enhanced. Such resources as wealth, status, size, weapons, intelligence, knowledge, organizational skill, internal unity, respect, affection, allies, and a reputation of being powerful are some of the bases of power. Effective power depends not only on the control or possession of resources to generate power but also upon the motivation to employ these resources to influence others, skill in converting these resources to usable power, and good judgment in employing this power so that its use is appropriate in type and magnitude to the situation in which it is used.

It is evident that a group's possession of highly effective power increases its chances of getting what it desires. Therefore, one would expect that the members of high-power groups would be more satisfied with their groups and more intent on preserving the status quo than would members of low-power groups. Given this asymmetry in power and satisfaction, it also could be expected that pressures for change in the power relations are most apt to come from low-power groups. The question naturally arises: How do high-power groups use their power to prevent or contain such pressure from low-power groups?

There are several basic ways: *control over the instruments of systematic terror* and of their use; *control over the state*, which establishes and enforces the laws, rules, and procedures which regulate the social institutions of the society; *control over the institutions* (such as the family, school church, and media) *which socialize and indoctrinate people* to accept the power inequalities; and *interactive power*, in which repeated individual behaviors by those who are more powerful confirm the subordinate status of those in low power. In addition, there are the *self-fulfilling prophecies* in which the behavior of the oppressed, resulting from their oppression, are used by the oppressor to justify the oppression; and the *distorted relation* between the oppressor and the oppressed.

4.5.2 Systematic Terror

As Sidanius/Pratto (1999: 41) point out in their excellent book *Social Dominance*, systematic terror can be official, semi-official, or unofficial. "*Official terror* is the public and legally sanctioned violence and threat of violence by organs of the state toward members of a subordinate group" (as in the behavior of South African police toward Blacks during the Apartheid period). *Semi-official terror* is violence or intimidation carried out by officials of the state but not legally sanctioned by the state (e.g., the death squads in Argentina composed of paramilitary organizations), while *unofficial terror* is perpetrated by private individuals from dominant groups, often illegally, with the tacit approval of public officials (as in the lynchings of African-American men accused of having sex with White women).

Systematic terror may not be necessary to keep a subordinated group in its place, if they think the social institutions controlled by the dominant group, as well as their daily interactions with its members, are tolerable. Or, it might be that their socialization and indoctrination by the social institutions controlled by the powerful have led them to accept and internalize the values and ideology of the dominant group. Even so, a harsh, dominant group in a totalitarian society may find it expedient, as well as self-affirming, to keep salient the potential of systematic terror, through its occasional arbitrary use to encourage the continued internalization of its values by the subordinate group and the toleration of the injustices it is experiencing.

4.5.3 Control over the State

In a self-reinforcing cycle, the powerful in any society control the state, and control of the state increases the power of those who control it. In the United States and other Western democracies, large corporations and wealthy individuals are the primary funders of political campaigns, political parties, and political candidates; they also own and control most of the mass media. Additionally, they provide the support for most of the private policy-planning network—the think tanks, research institutes, policy discussion groups, and foundations—that help to set the national policy agenda and to establish policy priorities (see Perrucci/Wysong 1999, Chaps. 4 and 5 for a detailed discussion). The result of the foregoing is an immense bias in the political system favoring large corporations and the economically privileged in the legislative, executive, and judicial branches of the government. The effects of this bias are evidenced in which groups experience the various forms of injustice described earlier in this paper. In the United States, it is apparent that such minorities as African-Americans, Hispanics, Native Americans, the physically impaired, single mothers, and children have relatively little power and are more likely to be poor and to suffer the other injustices associated with poverty. At the global level, a similar process occurs: the large multi-national corporations, the more powerful nations, and wealthy investors are able to influence the processes and practices affecting international trade, aid, and investment to their own advantage and often to the detriment of the people in third-world states.

4.5.4 Control over Socialization and Indoctrination

The development of discontent among the disadvantaged and outrage among the oppressed are often aborted by the socialization and indoctrination institutions of society. The family, school, religious institutions, and media socialize and indoctrinate the oppressed to obey authority and be aware that punishment for disobedience will be severe, to view the disadvantages they suffer as legitimate, or to have faith that they will be compensated for them in the afterlife. The rewards and punishments for accepting or challenging authority and the status quo in the here-and now, as well as in the afterlife, are presented vividly and repeatedly in both the myths and practices of the society and its indoctrinating institutions.

4.5.5 *Interactive Power*

This form of power has been defined by Harvey (1999: 43) as "the power to take the initiative in a relationship: in beginning or ending a relationship, and in insisting on its being modified, and in taking a number of communication initiatives like the power to begin or end a specific contact (like a conversation), to insist on being listened to and on being given answer to reasonable and pertinent questions." The socially privileged, typically, assume that they have the right to control the interactions in their relationship with members of subordinated groups. Challenging this assumption can be risky for a subordinate and, as a consequence, it usually goes unchallenged. The repeated, everyday experience of being treated as an inferior produces a public image of being an inferior, which may be internalized as an image of self-inferiority. In the socially privileged, in contrast, such interactions will produce a public image of superiority and a corresponding self-image. Such non-egalitarian everyday interactions between the socially dominant and the oppressed help to keep the system of oppression in place through the public images and self-images they produce and perpetuate.

4.5.6 *The Social Production of Meaning in the Service of Legitimating Oppression*

Under this heading, we will provide some illustrations of how the various institutions of society and facets of its culture implicitly "proclaim the superiority of the oppressor's identity" (Noel 1994: 7). The oppressors use 'history,' "the law of nature," "the will of God," 'science,' "the criteria of art," and 'language', as well as the social institutions of society, to legitimize their superiority and to ignore or minimize the identity of the oppressed.

Some illustrations (see Noel 1994, for a more detailed discussion) follow:

The Declaration of Independence starts with "We the People" but the 'we' did not include Native Americans, slaves, women, or youth.

- 'History,' as it appears in the textbooks, is mainly a series of events that involve "great men" such as conquerors, kings, presidents, or successful revolutionary leaders. They were the 'winners;' the losers, if mentioned, are usually presented in a derogatory manner. The history of women, African-Americans, Native Americans, children, the aged, homosexuals, the physically challenged, and other minority groups are too insignificant to be noted except as problems.

- Pseudoscientific "Social Darwinism" eagerly misapplied such ideas as survival of the fittest, hereditary determinism, and stages of evolution to the relations between different human social groups—classes and nations as well as social races—in order to justify existing exploitative social relations and to rationalize imperialist policies. "The rich and powerful were biologically superior; they had achieved their positions as a result of natural selection. It would be against nature to interfere with the inequality and suffering of the poor and weak. Imperialism was patriotism in a race endowed with the genius for empire, for those superior peoples meant to lead inferior peoples" (Deutsch 1973: 102–103).
- All the large-scale religions share the belief in female inferiority (Noel 1994). God, according to the Christian tradition, made man in *his* own image, while a woman is a mere reflection of man. In Hinduism, women are not even eligible for salvation; they must await another incarnation. In Islam, the testimony of a woman is worth only half that of a man. Everyday, the Orthodox Jewish male thanks God "for not having made him a woman." According to Pope John Paul II, women are not allowed to be priests because this would be contrary to both their humanity and femininity.
- The behavioral and social sciences have often legitimized the oppressors' claim to superiority. Well-known psychologists have used the results of intelligence testing to proclaim that African-Americans, Jews, Eastern Europeans, and people from the Mediterranean area are inferior to Anglo-Saxons. Piaget and Kohlberg indicated that women have a less developed moral judgment than men. Sociologists (e.g., Banfield) have considered the lower classes to be pathological, anthropologists have employed the term 'primitive' to characterize indigenous societies, and psychiatrists have considered homosexuality to be a mental disease, women to suffer from "penis envy," and children to fantasize their abuse.
  - The historians of art, music, and literature have much neglected the contributions of women and have frequently credited their works to men; 'art' and 'literature' are created by the dominators. African art is 'primitive' art, even though copied by Picasso, 'gays' write 'homosexual' novels, and female film directors produce 'women's' movies. It has long been accepted for minority artists and performers to work in their own group's genre—for example, for Blacks to create and perform jazz music. Only recently have Blacks been permitted to express themselves in the "higher genres" of classical music, ballet, or opera.

4.5.7 The Contribution of Self-fulfilling Prophecies to the Maintenance of Oppression

The myths of the moral, intellectual, or motivational superiority of the oppressor, which often are used to legitimize the subordination of oppressed groups, are typically supported by self-fulfilling prophecies. As Sidanius/Pratto (1999: 227) point out: "Societies are set up in ways that make life relatively easy for dominants and relatively difficult for subordinates." Subordinated groups are less likely to live in circumstances that encourage and stimulate the development of one's intellectual potential, foster the motivation to be ambitious and to achieve economic success, motivate conformity to the social norms against deviant and criminal behavior, foster intragroup cohesiveness, and contribute to the development of physical and mental health. These deficiencies resulting from oppression support the mythology and stereotypes promulgated by the oppressor and, in a self-fulfilling prophecy, enable the dominant to justify their oppression by characterizing the oppressed as being "dumb, lazy, or immoral".

Of course, there are oppressed groups who do not fit these stereotypes. Such groups, which have high intellectual and economic attainments as well as much intragroup cohesiveness, are often viewed as potential competitors. They are stereotyped as "cunning, deceitful, overly ambitious, and clannish". These groups tend to be morally excluded or marginalized so that they have only restricted or limited participation in the important institutions of society: political, legal, educational, etc. They tend to be segregated from the dominant group, their economic activities are primarily in stigmatized occupations, and often they have to be very ambitious, cunning, and clannish to survive and thrive. As I have suggested elsewhere (Deutsch/Collins 1951), these groups are seen as potential competitive threats to the dominant group, and the responses to such threats often take the forms of intolerance, exclusion, or extermination.

4.5.8 The Distorted Relationship Between the Oppressed and the Oppressor

4.5.8.1 The Oppressed

Imagine the situation of an oppressed or abused child, wife, employee, or citizen. Each is in some critical way dependent upon the oppressor—the parent, the husband, the employer (company or organization), and the governing undemocratic power. Suppose the oppressed has needs or desires of which the oppressor strongly disapproves (e.g., physical affection, self-esteem, autonomy, self-determination) or only allows to be expressed in distorted dissatisfying, self-abusive forms. The reaction of the oppressed is apt to be one of frustration → anger → anxiety if the oppressor indicates, even subtly, that the oppressed will be severely punished if she expresses her desires, frustrations, or anger. One way of reducing the anxiety aroused by temptations to manifest the forbidden desires is to build an internal barrier to their expression by internalizing the threat through *identification with the oppressor* (Freud 1937). Doing so leads, at one level, to guilt and self-hatred for having these desires. At a deeper level, it leads to guilt and self-hatred for abandoning one's self, as well as to rage and a sense of moral superiority toward the oppressor, who is responsible for this abandonment. As a result of these processes, submission and obedience to the oppressor, as well as depression, are commonly found among the oppressed when they are interacting with oppressors or when they are in oppressive situations.

However, it should be recognized that many who experience oppression in some aspects of their lives do not necessarily experience it in other aspects, so that they are not necessarily submissive and depressed personalities racked by guilt, self-hatred, and rage in all situations. Damage to the personalities of oppressed people will be limited, even when exposed to pervasive oppression, if they are also part of a supportive, cohesive community whose values oppose oppression.

4.5.8.2 The Oppressor

If we were to examine the oppressors psychologically—child abusers, husbands who batter their wives, brutal bosses, and political tyrants—I believe that we would find that the oppressors need the oppressed. Their need to control and dominate the other; their intolerance of the autonomy of the other makes them dependent upon having vulnerable, weaker others for the definition of their own power. Their own deep sense of vulnerability (anxieties about helplessness and impotence, guilt about forbidden desires and rage, self-hatred for vulnerability) leads to strong needs both to deny their vulnerability (by projecting their anxieties, guilt, and contempt onto others who are more vulnerable) and to have the power to control those who are vulnerable or can be made to be more vulnerable. The oppressor needs to be able to make arbitrary and unreasonable demands so that the obedience of the oppressed is due to the oppressor's power and not to the agreement of the oppressed. The oppressor's intolerance of the autonomy of the oppressed is (Lichtenberg 1990: 26) "neither idle nor freely chosen; it is a function of dependence on the vulnerable others for the definition of his or her own power".[3]

One can, of course, be more powerful in a relationship (such as a parent-child, employer-employee relationship) without being an oppressor. Power can be used 'for' the other rather than 'against' the other.

4.5.8.3 The Psychodynamic Relationship of the Oppressor and the Oppressed

There are structural similarities between the sadomasochistic and the oppressor-oppressed relationship. Each side of the relationship has some of the latent qualities of the other side: the sadist when he is whipping the masochist is also whipping

[3] Also, it should be noted that the social science literature on "quality-of-life" and "subjective well-being" indicates that there is a low contribution, beyond the poverty level, of greater income to subjective well-being in advanced economies (Kahneman et al. 1999; Lane 2000). The oppressors' drive for superior wealth, accompanied by conspicuous consumption, often impairs social relations, a key component of subjective well-being.

himself; the oppressor when he is controlling the oppressed is controlling himself. The masochistic, when whipped, is also having the sadist within himself punished. Similarly, the oppressed who is being controlled is also having his rage controlled.

It seems obvious that not all oppressors have 'oppressive' personalities nor do all the oppressed have 'oppressed' personalities, in the sense that they do not consistently prefer and seek out relationships where they can be the 'oppressors' or the 'oppressed.' Nevertheless, I suggest that in any longstanding oppressive relationship, both the psychodynamics within its participants as well as social expectations will contribute to its persistence and resistance to change. Thus, in Afghanistan, despite the ending of the Taliban's rule and their exposure to different models of family relationships on TV, many wives will continue to believe that their husbands have the right to beat them if they disobey them.

I conclude this section of my discussion by stating that any attempt to end long-enduring oppressive relations will have to address the psychodynamic issues that lead people to resist changing unhappy but familiar relationships. Some of the anxieties and fears that have to be addressed for the oppressed and the oppressor are listed as follows:

1. Both feel anxious in the face of the unknown. They believe that they will be foolish, humiliated, or helpless in a new, unclear relationship.
2. Both fear the guilt and self-contempt for their roles in maintaining the oppressive relationship.
3. The oppressed fear that their rage will be unleashed; the oppressor is in terror of this rage.
4. Both fear punishment if they change: the oppressed from the oppressor, the oppressor from the oppressed and other oppressors.
5. Both anticipate loss from the change: the oppressed will lose their sense of moral superiority and the excuses of victimhood; the oppressor will lose the respect and material benefits associated with being more powerful.

4.6 Awakening the Sense of Injustice

In this section, I shall consider the sensitivity to injustice in the victim and the victimizer. Awareness of injustice is a precondition for overcoming it. For a more extended discussion see Deutsch/Steil (1988).

4.6.1 The Differential Sensitivity to Injustice of the Victim and the Victimizer

Although it may be morally better "to be sinned against than to sin," it is generally accepted that the immediate pain is usually greater for the one who is sinned against than for the sinner. As I have indicated earlier, the victimizers—in addition to their gains from their exploitative actions—commonly have the reassurance of official definitions of justice and the support of such major institutions as the church, the press, and the schools to deaden their sensitivities to the injustices inherent in their relations with the victim. The victim may, of course, be taken in by the official definitions and the indoctrination emanating from social institutions and, as a result, lose his sensitivity to injustice. Because he is the one who is experiencing the negative consequences of the injustice, he is also less likely to feel committed to the official definitions and indoctrinations because of his lack of participation in creating them.

The explanation for differential sensitivity in terms of differential gains and differential power is not the complete story. There are, of course, relations in which the victimizer is not of superior power, and yet, even so, he will not experience guilt for his actions. Consider a traffic accident in which a car hits a pedestrian. The driver of the car will often perceive the accident so as to place responsibility for it upon the victim. Seeing the victim as responsible will enable the driver to maintain a positive image of himself. Projecting the blame onto the victim enables the victimizer to feel blameless.

If we accept the notion that most people try to maintain a positive conception of themselves, we can expect a differential sensitivity to injustice in those who experience pain, harm, or misfortune and those who cause it. If I try to think well of myself, I shall minimize my responsibility for any injustice that is connected with me or minimize the amount of injustice that has occurred if I cannot minimize my responsibility. On the other hand, if I am the victim of pain and harm, to think well of myself, it is necessary for me to believe that it was not my due: it was not a just dessert for a person of my good character. Thus, the need to maintain positive self-esteem leads to opposite reactions in those who have caused an injustice and those who suffer from it.

Although the need to maintain a positive self-regard is common, it is not universal. The victim of injustice, if he views himself favorably, may be outraged by his experience and attempt to undo it. In the process of doing so, he may have to challenge the victimizer. If the victimizer is more powerful and has the support of the legal institutions and other institutions of the society, the victim will realize that it would be dangerous to act on his outrage or even to express it. As I have indicated earlier, under such circumstances, in a process that Anna Freud (1937) labeled "identification with the aggressor," the victim may control his dangerous feelings of injustice by denying them and by internalizing the derogatory attitudes that the victimizer has toward him.

4.6.2 Conditions that Awaken and Intensify the Sensitivity to Injustice

In the preceding, I have suggested that the sense of injustice may be minimal in the oppressors and also in the oppressed under certain circumstances. Here, I wish to consider the conditions that awaken and intensify the sensitivity to injustice.

The major explanatory theme advanced by social scientists for the sensitivity to injustice is that of *relative deprivation*. This is the perceived discrepancy between what a person believes she is entitled to and what she obtains, regarding the different forms of justice described in the third section: distributive justice, procedural justice, retribution, moral inclusion, and cultural imperialism. It is commonly assumed that it is relative rather than absolute deprivation that is critical in stimulating dissatisfaction. Research (see Crosby 1982, for an excellent summary) has demonstrated that people who are well off by absolute standards may feel more discontent than those who are much worse off if they feel relatively more deprived because their aspirations are high or because they are surrounded by people who are even more well-off than they are.

Runciman (1966) made a distinction between two types of relative deprivation: egoistic and fraternal. *Egoistical* deprivation occurs when an individual feels disadvantaged relative to other individuals; *fraternal* deprivation occurs when a person feels his group is disadvantaged in relation to another group. An individual may feel doubly deprived, as an individual and as a group member. As Tajfel (1982) pointed out, the two kinds of deprivations have different implications for how an individual may improve his situation. To remedy fraternal deprivation, social change (change in the position of one's group) is necessary. To remedy egoistic deprivation only entails change in one's individual situation.

The greater the magnitude of relative deprivation, the greater the sense of injustice that will be experienced by the oppressed. Members of the relatively advantaged group will be sensitive to the injustices experienced by the oppressed when they are aware that the oppressed are relatively deprived and that they are receiving less than their entitlement.

An individual's conception of what he and others are entitled to is determined by at least five major kinds of influence: (1) the ideologies and myths about justice that are dominant and officially supported in the society, (2) the amount of exposure to ideologies and myths that conflict with those that are officially supported and are supportive of larger claims for the oppressed, (3) experienced changes in satisfaction-dissatisfaction, (4) knowledge of what others who are viewed as comparable are getting, and (5) perceptions of the bargaining power of the oppressed and oppressors.

4.6.2.1 The Influence of Ideologies and Myths

The official ideology and myths of any society help define and justify the values that are distributed to different positions within the society; they codify for the individual what a person in his position can legitimately expect. Examples are legion of how official ideology and myth limit or enhance one's views of what one is entitled to. The American poor offer an instance of the potency of myth in creating an identity that promotes docility in the face of deprivation (Edelman 1971). Americans are taught by their schools, the mass media, and their political rhetoric that America is the land of equal opportunity. Given such pervasive indoctrination, the poor are apt to attribute their condition to their own failings. This view of themselves as unworthy is further supported by cues from governmental practices toward them which place in question their morality, ambition, and competence. As a result, the poor in America have typically been meek and acquiescent, requiring less coercion and less in benefits than has been true in other developed countries. Similarly, the ideology and myth of White supremacy has led Whites to expect that they are entitled to deferential behavior from Blacks and Blacks to expect that they are not entitled to equal treatment from Whites. Similarly, men and women under the influence of a sexist mythology and ideology have defined gender entitlements that give the women supremacy in the narrow confines of the kitchen and the nursery, while men have supremacy in the broad world outside the home as well as in many areas within it.

4.6.2.2 The Weakening of Official Ideologies

It is difficult not to accept the official myths and ideology of one's society, even if they are to one's disadvantage, unless (1) there is a breakdown of consensual norms and the inability or unwillingness of the ruling elite to act in such a way as to restore these norms; this is likely to occur during a period of rapid social change or intra-societal conflict, either of which could bring into question the legitimacy of traditional myths and values; (2) there is a failure of the society to deliver the entitlements that it has defined as legitimate for one's position so that the magnitude of one's relative deprivation is increased; this could be due to natural or social disasters that worsen the conditions of daily life; or (3) there is exposure to new ideologies and new examples that are accepted as legitimate by many people, which stimulate consciousness of new and better possibilities. This could happen as the result of increased communication arising from new technological developments such as books, newspapers, radio, and television, or it may reflect increased urbanization and the resulting exposure to more diversity of people, ideas, and experience. Obviously, one would expect that the receptivity to new ideologies and examples would be heightened by the breakdown of legitimacy of the existing ideology and the worsening of living conditions.

4.6.2.3 Experienced Changes in Satisfactions-Dissatisfactions

Modifications in the conception of what one is entitled to derive not only from alterations in the ideology and myths that one accepts but also from changes in one's experiences of satisfaction. A period of gain creates expectations about further improvement. As de Tocqueville commented in *L'Ancien Regime* (1947): "Only a great genius can save a prince who undertakes to relieve his subjects after a long oppression. The evil, which was suffered patiently as inevitable, seems unendurable as soon as the idea of escaping from it is conceived" (p. 186).

Many social scientists, before and after de Tocqueville, have written insightfully about the "revolution of rising expectations" to explain the paradox that social discontent and even revolutionary activity are more likely to occur after social conditions have improved; when there is rising hope, not bleak despair. The explanation generally follows two major lines. First, improvement of social conditions increases aspirations by increasing what is perceived to be possible to attain. Demand may increase at a faster rate than the actual gains received, with a resulting increase in relative deprivation and in the sense of injustice. The increased discontent is most likely to occur if the gains are discontinued or reversed after the initial gains have heightened further expectations.

The second explanation of the effects of gains is that the increase is not uniform in all areas in which the victimized are disadvantaged. Improvement in one area, such as education, only makes one more sensitive to the injustice one is experiencing in other areas, such as employment, police protection, and housing. Many social scientists have advanced the proposition that status disequilibrium (such that there are differences in one's relative statuses in income, education, social prestige, and the like) is a source of tension and discontent (e.g., Davies 1962; Galtung 1964; Himmelstrand 1969). Thus, a very effective way of enhancing the sense of injustice of the victimized is to increase their education and little else.

4.6.2.4 Comparing Oneself to Others

Alterations in the conception of what one is entitled to result not only from changes in the level of satisfaction but also from modifications to one's views, either about how comparable others are being treated or about who should be considered as comparable. There is considerable research evidence that one's attitude, one's evaluations of one's abilities, and one's emotions are very much influenced by one's perceptions of these attributes in others who are used for comparison purposes (see Pettigrew 1967, for summary). Although the evidence is by no means conclusive, it has been suggested by Festinger (1954), Gurr (1970), and others that comparison tends to be primarily with similar others, and Gurr further suggests that the comparison will be with the similar others whose gains are most rapid. Thus, if someone else who is perceived to be similar is already better off, then one will feel that it is unjust. If, in addition, the person advances rapidly in status, salary, or the like, one will experience a substantial increase in relative deprivation unless one receives a comparable increase. A potent way of arousing the sense of injustice is to make the victim more aware that comparable others are being treated better or to increase her feeling that it is appropriate to compare herself to others whom she previously considered to be incomparable to herself.

4.6.2.5 Increasing Bargaining Power

One's perceived power is undoubtedly a factor determining what one is entitled to and with whom one compares oneself to establish one's entitlements. If a victim or victimized group is dealing with an unresponsive exploitative group, it is faced with either the possibility of resigning into apathy and depression or the possibility of attempting to increase its power sufficiently to persuade or compel the other to negotiate. Bargaining power is increased by either of two means: increasing one's own power or decreasing the other's power. Attempts to change power can be directed at altering the resources that underlie power (such as wealth, physical strength, organization, knowledge, skill, trust, respect, and affection), or they can be directed toward modifying the effectiveness with which the resources of power are employed.

The primary resources of the oppressed are the number of discontented people and the fact that they have justice on their side. The utility of people as a resource for power is a function of their *numbers*, their *personal qualities* (such as their knowledge, skill, dedication, and discipline), their *social cohesion* (as reflected in mutual trust, mutual liking, mutual values, and mutual goals) and their *social organization* (as expressed in effective coordination and communication, division of labor, and specialization of function, planning, and evaluation). Numbers of people are obviously important but undoubtedly not as important as their personal qualities, social cohesion, and social organization. A large, inchoate mass of undisciplined, ineffectual people is at the mercy of a small, dedicated, disciplined, well-organized, cohesive group. Most large groups are controlled by less than 10 % of their membership.

If one examines such low-power minority groups as the Jews, Chinese, and Japanese, who have done disproportionately well in the United States and in other countries to which they have migrated, it is apparent that these groups have been characterized by high social cohesion and effective social organization, combined with an emphasis upon the development of such personal qualities as skill, dedication, and discipline. Similarly, the effectiveness of such guerilla forces as the Vietcong or such terrorist groups as Al Qaeda has been, in part, due to their cohesion, social organization, and personal dedication. Clearly, the development of these characteristics is of prime importance as a means of increasing the power of one's group.

Elsewhere (Deutsch 1973), I have considered some of the determinants of cohesion. Here I add that groups become cohesive by formulating and working together on issues that are specific, immediate, and realizable. They become effectively organized as they plan how to use their resources to achieve their purposes and as they evaluate their past effectiveness in light of their experiences. It is apparent that the pursuit of vague, far-in-the-future, grandiose objectives will not long sustain a group's cohesiveness. Nor will the exclusive pursuit of a single issue be likely to sustain a long-enduring group unless that issue proliferates into many sub-issues. Those intent upon developing social cohesion and social organization should initially seek out issues that permit significant victories quickly; they will set out on a protracted indeterminate struggle only after strongly cohesive and effective social organizations have been created.

So far, I have stressed personal qualities, social cohesion, and social organization as resources that can be developed by low-power groups to enhance their power. Typically, such resources are vastly underdeveloped in victimized groups; however, they are necessary for the effective utilization of almost every other type of resource, including money, votes, tools, force, and the like. Low-power groups often have two other key assets that can be used to amplify their other resources: discontent and the sense of injustice. If intense enough, these may provide the activating motivation and the continuing determination to change the status quo. They are the energizers for individual and social action to bring about change. Moreover, to the extent that the basis for discontent and the nature of the injustice can be communicated to others so that they experience it, if only vicariously, supporters and allies will be attracted to the side of the low-power group. Increasing the number of one's supporters and allies is another important way of increasing one's power. Thus, in a circular way, bargaining power and the sense of injustice mutually reinforce each other: an increase in one increases the other.

Discontent and the sense of injustice may be latent rather than manifest in a subordinated group. Neither the consciousness of oneself as victimized or disadvantaged nor the consciousness of being a member of a disadvantaged class may exist psychologically. If this is the case, consciousness-raising tactics are necessary precursors to the developing of group cohesion and social organization. The diversity of consciousness-raising tactics has been illustrated by the variety of techniques employed in recent years by Black power groups, women's liberation groups, and gay and lesbian groups. They range from quasi-therapeutic group discussion meetings to mass meetings and demonstrations to dramatic confrontations of those in high-power groups. It is likely that a positive consciousness of one's disadvantaged identity is most aroused when one sees someone who is considered to be similar to oneself explicitly attacked or disadvantaged and sees him resist successfully or overcome the attack; his resistance reveals simultaneously the wound and its cure.

4.7 Overcoming Oppression

Once the sense of injustice has been awakened, to change the oppressive relation there are basic strategies available: persuasion and power strategies. Each is discussed in the following sections.

4.7.1 Persuasion Strategies

These strategies are aimed at convincing those in high power to change so that power is shared more equitably and oppressive practices are reduced or eliminated. There are three main types: *appeals to moral values*, *appeals to self-interest*, and *appeals to self-realization.* For excellent discussions of the psychological processes involved in persuasion in conflict and negotiating situations, see Chaiken et al. (2000). For a systematic discussion of the multiple factors affecting attitude change, see Petty and Wegener (1998).

4.7.1.1 Appeals to Moral Values

These assume that the oppressor is not fully aware of the unjust situation of the oppressed and that if he were so, his conscience or moral values would move him to take action to remedy the situation. The appeals are aimed at both cognitions and affect so that the oppressor can understand how his moral values are being violated by the injustices and can feel sufficiently guilty or outraged to take action to eliminate the injustices. This sort of empathic understanding of the injustices experienced by various subordinated groups can be developed in many ways. The most effective way is by experiencing, directly or indirectly, what it is like to suffer the injustices. Indirect experiences include conversations with members of an oppressed group about their life experiences, tutored role-playing as a member of such groups; reading autobiographies and novels, watching films and videos that dramatize and make emotionally vivid the experience of injustices; and hearing lectures and sermons which make salient the moral values being violated.

If the oppressor believes that he has the moral right to engage in oppressive practices (e.g., beating his wife when she disobeys him), then attempts to create empathic understanding of the situation of the oppressed is not likely to be successful. Here, what is needed is a moral authority (e.g., the legal system, religious authority, the consensus of his peers), which he accepts as superior to his own sense of morality, to persuade him that he is morally wrong. However, unfortunately, in many situations the powerful are not responsive to moral persuasion because the moral authority endorses the oppression or the oppressor is indifferent to moral claims.

4.7.1.2 Appeals to Self-interest

Such appeals are often more effective for people who are embedded in an individualistic or competitive society. In such cases, the process of persuasion starts with the communicator having a message that he wants to get across to the other. He must have an objective if he is to be able to articulate a clear and compelling message. Further, in formulating and communicating his message, it is important to recognize that it will be heard not only by the other but also by one's own group and by other interested audiences. The desirable effects of a message on its intended audience may be negated by its unanticipated effects on those for whom it was not intended.

I suggest that, to be effective, the oppressed Acme's message to Bolt (the oppressor) should include the following elements:

(1) A clear statement of the specific actions and changes being requested of Bolt. Bolt should know what is expected of him so that he can fulfill Acme's expectations if he so desires. Presumably, Bolt is more apt to do what Acme wishes if Bolt believes that it is possible for him to do so. He is more likely to believe that this is the case if Acme's wants are perceived to be specific and limited, rather than if they are viewed as vague and unbounded.

(2) An appreciation of the difficulties, problems, and costs that Bolt anticipates if he complies with Acme's wishes. Such an appreciation should be combined with an expressed willingness to cooperate with Bolt to overcome the difficulties and reduce the costs. This willingness entails a readiness on Acme's part to consider Bolt's proposals and counter proposals and to modify his own initial proposals so that a mutually responsive agreement can be reached.

(3) A depiction of the values and benefits that Bolt will realize by cooperating with Acme. In effect, if Bolt can be persuaded that he has more to gain than to lose by doing what Acme wants, obviously, he is more likely to do it. The important gains reside in the possibility that Bolt, sharing power with Acme, may enhance Acme's general cooperativeness and thus markedly increase Bolt's fulfillment of his own objectives. There are many instances in labor-management, student-faculty, and warden-prisoner relations that indicate that the more powerful party has gained enormously through enhancing the power (and thus the sense of responsibility) of the weaker party. In addition, other dissatisfactions that Bolt has experienced in his relationship with Acme may be reduced by Acme's enhanced cooperativeness. Other sources of potential gain for Bolt reside in the enhanced reputation and goodwill that he will obtain from influential third parties and in the greater fulfillment he will experience when Acme is content rather than dissatisfied with their relationship.

(4) A statement of the negative, harmful consequences that are inevitable for Bolt's values and objectives if Acme's wishes are not responded to positively. In effect, Bolt has to be led to understand the costs of nonagreement so that he can realize that the costs of agreement are not the only costs to be taken into account. Potential costs for Bolt of a failure to come to an agreement include: the losses resulting from a decrease in Acme's future cooperativeness, including the possibility of Acme's total noncooperation, losses in esteem and goodwill, possibly the loss of cooperation of significant third parties, and losses due to active attempts to embarrass, harass, obstruct, or destroy the interests of Bolt by Acme or by his sympathizers.

(5) An expression of the power and resolve of Acme to act effectively and unwaveringly to induce Bolt to come to an acceptable agreement. Acme's unshakable commitment to induce a change may affect Bolt by convincing him that Acme's needs are serious rather than whimsical and thus deserve fulfillment. It may also persuade Bolt that the pressure from Acme will not diminish until an acceptable agreement has been reached. However, if Bolt has no concern whatsoever for Acme's needs and no belief that Acme's pressure will be sufficiently strong to be disturbing, Acme must attempt to develop, mobilize, and publicize its power sufficiently to convince Bolt that negotiation would be a prudent course of action.

A message that contains the elements mentioned earlier strongly commits Acme to his objective yet suggests that the means of attaining it are flexible and potentially responsive to Bolt's views. Because the objective is articulated so as to be specific and limited, it is more likely to be considered by Bolt as feasible for him to accept than one stated in more generalized and grandiose terms. The message provides Bolt with the positive prospect that changes will result in enhanced social and self-esteem and that they will yield the benefits to be derived from increased cooperation from Acme. It also indicates the negative results to be expected from lack of change. Although Acme's firm intent to alter the status quo is made evident, his stance throughout is cooperative. The possibility of a true mutual exchange is kept open, with explicit recognition that the dissatisfactions and the problems are not one-sided.

4.7.1.3 Appeals to Self-realization

These are also involved in appeals to self-interest. Here, I am more specifically referring to the distortions of self that are involved in the distorted relationship between the oppressor and the oppressed. As Lichtenberg (1990: 191–192) asks: "If the rich are doing so well why aren't they happy? Why is there so much alcoholism among the power elite, so much drunkenness, so much attachment to non-essentials, like 'pinstripes on one's Mercedes'?"

For the oppressor to attain an undistorted self, Lichtenberg (1990) suggests that, not only must he withdraw from the processes of domination, he must re-own and resolve his feelings of vulnerability, guilt, and self-hatred and his rage and terror. In addition, he must undo the projection of these feelings onto the oppressed. How can the oppressor be helped to this self-realization? Psychologists, in their roles as psychotherapists, marriage counselors, organizational consultants, and educators, have a role to play in demystifying the psychological processes involved in the dominators. I believe that the oppressed also have a role to play, by not accepting their distorted roles in the distorted relationship of the oppressor and the oppressed.

4.7.2 Difficulties That Interfere with the Use of Persuasive Messages by Low-Power Groups

Rage or fear in the low-power group often makes it impossible for the members of that group to communicate persuasive messages of the sort described earlier. Rage leads to an emphasis on destructive, coercive techniques and precludes offers of authentic cooperation. Fear, on the other hand, weakens the commitment to the steps necessary to induce a change and lessens credibility regarding the idea that compliance will be withdrawn if change does not occur. Rage is potentially a more useful emotion than fear, because it leads to actions that are less damaging to the development of a sense of power and, hence, of self-esteem. Harnessed rage or outrage can be a powerful energizer for determined action, and if this action is directed toward building one's own power rather than destroying the other's power, the outrage may have a socially constructive outcome.

In any case, it is evident that when intense rage or fear is the dominant emotion, the cooperative message outlined here is largely irrelevant. Both rage and fear are rooted in a sense of helplessness and powerlessness; they are emotions associated with a state of dependency. Those in low power can overcome these debilitating emotions by their own successful action on matters of significance to them. In the current slang, they have to "do their own thing"; it cannot be given to them or done for them. This is why my emphasis is on the sharing of power, and thus increasing one's power to affect one's fate, rather than on the sharing of affluence. While the sharing of affluence is desirable, it is not sufficient. In its most debilitating sense, poverty is a lack of power and not merely a lack of money. Money is, of course, a base for power, but it is not the only one. If one chooses to be poor, as some members of religious or pioneering groups do, the psychological syndrome usually associated with imposed poverty—a mixture of humiliation, dependency, victimhood, apathy, small time-perspective, suspicion, fear, and rage—is not present.

Thus, the ability to offer and engage in authentic cooperation presupposes an awareness that one is neither helpless nor powerless, even though one is at a relative disadvantage. Not only independent action but also cooperative action requires a recognition and confirmation of one's capacity to "go it alone" if necessary. Unless one has the freedom to choose *not* to cooperate, there can be no free choice *to* cooperate. Powerlessness and the associated lack of self and group esteem are not conducive to either internal group cohesiveness or external cooperation. Power does not, however, necessarily lead to cooperation. This is partly because, in its origin and rhetoric, power of the oppressed group may be oriented against the power of the established and thus likely to intensify the defensiveness of those with high power.

However, even if power is 'for' rather than 'against,' and even if it provides a basis for authentic cooperation, cooperation may not occur because it is of little importance to the high-power group. This group may be unaffected by the positive or negative incentives that the low-power group controls; it does not need their compliance. Universities can obtain new students, the affluent nations are no longer so dependent upon the raw materials produced in the underdeveloped nations, and the White industrial society does not need many unskilled Black workers.

4.7.3 Power Strategies

Apart from resigning into depression, what can members of a low-power group do when the dominant group is unwilling to negotiate a change in the status quo? Basically, there is only the possibility of increasing the group's relative power sufficiently to compel the other to negotiate. Relative power is increased by either of two means: enhancing one's own power or decreasing the other's power.

4.7.3.1 Enhancing One's Own Power

As I have indicated earlier, it involves increasing one's possession of the resources on which power is based and increasing the effectiveness with which the power is used. There are three areas in which those with low power can find additional resources: *within one's self or group*; *within potential allies*; and *within the oppressor.* Mandela (1994) in his autobiographical book, *Long Walk to Freedom*, provides many illustrations of how he did this, even when he was a prisoner of the repressive, apartheid South African government.

4.7.3.2 Developing Power Within One's Self or One's Group

By exerting considerable self-discipline while he was a prisoner, Mandela kept himself in excellent physical and mental condition. He stated that when he was a prisoner on Robben Island, the notorious prison island (Mandela 1994: 427): "On Monday through Thursday, I would do stationary running in my cell in the morning for up to 45 min. I would also perform one hundred fingertip push-ups, two hundred sit-ups, fifty deep knee-bends and various other calisthenics." He kept himself in good shape mentally by reading widely, by becoming an informed expert on the laws and regulations concerning the treatment of prisoners, and by studying for an L.L.B. degree at the University of London.

And he kept his self-concept undistorted by preserving his dignity and refusing to submit, psychologically, to the definition of self that the oppressors tried to force upon him. For example, he described the following incident after landing on Robben Island (Mandela 1995: 297–299):

> We were met by a group of burly white wardens shouting: "Dis die Eiland! Hier gaan jiell vrek! (This is the island! Here you will die!)… As we walked toward the prison, the guards shouted 'Two-two! Two—two!'—meaning we should walk in pairs… I linked up with Tefu. The guards started screaming, 'Haas!… Haas!' The word haas means 'move' in Afrikaans, but it is commonly reserved for cattle.
> The wardens were demanding that we jog, and I turned to Tefu and under my breath said that we must set an example; if we give in now we would be at their mercy…
> I mentioned to Tefu that we should walk in front, and we took the lead. Once in front, we actually decreased the pace, walking slowly and deliberately. The guards were incredulous (and said)… we will tolerate no insubordination here. Haas! Haas! But we continued at our stately pace. (The head guard) ordered us to halt and stood in front of us: Look, man, we will kill you, we are not fooling around… This the last warning. Haas! Haas!
> To this I said: 'You have your duty and we have ours.' I was determined that we would not give in, and we did not, for we were already at the cells.

By his persistent public refusal to be humiliated or to feel humiliated, Mandela rejected the distorted, self-debilitating relationship that the oppressor sought to impose upon him. Doing so enhanced his leadership among his fellow political prisoners and the respect he was accorded by the less sadistic guards and wardens of the prison.

4.7.3.3 Allies Are Very Important

The acquisition of allies is central to enhancing the power of oppressed groups. Leaders of the African National Congress (ANC) of South Africa devoted considerable effort to developing allies among the leaders of other African nations as well as among many other influential groups in the UN, the Commonwealth, and the various industrial nations with economic ties to South Africa. There is little doubt that the allies they developed played a crucial role in bringing about the ending of the apartheid system and the formation of a new government with Nelson Mandela as President and the ANC as the dominant political party. Their allies did this by bringing sufficient economic, political, and moral pressure upon the apartheid government to convince the economic leaders of the country that a change was necessary if they were to avoid an economic disaster.

Unfortunately, oppressed groups sometimes do not sufficiently realize the important potential for allies among other oppressed groups. They may narrowly define their interests as overcoming the injustices that they are experiencing and may not be concerned with those being suffered by other oppressed groups. In the United States, for example, there is not an effective working coalition among such oppressed groups as Blacks, gays, women, Hispanics, the disabled, the poor, and the elderly because these separate groups do not define their interests inclusively. While every group has to be for itself, when it is also for others, it becomes stronger from the support it receives.

4.7.3.4 The Oppressor's Power Can Often Be Used Against the Oppressor by the Oppressed

As Alinsky (1971: 152) indicates: "Since the Haves publicly pose as the custodians of responsibility, morality, law and justice (which are frequently strangers to each other), they can be constantly pushed to live up to their own book of morality and regulations. No organization, including religion, can live up to the letter of its own book."

Alinsky (1971) cites many examples of tactics in which bureaucratic systems were snarled in their own red tape by pressure to live up to their own formally stated rules and procedures. Tactics of this sort may center upon demanding or using a service that one is entitled to that is not ordinarily used so massively, and for which the institution is not prepared to provide in large volume without excessive cost to itself. For example, banks may be disrupted by a massive opening and closing of accounts, department stores by massive returns of purchases, airports by a massive use of their toilets and urinals by visitors, and so forth. Or, the tactics may center upon disobedience to a rule or law that cannot be enforced in the face of massive noncompliance. Thus, landlords cannot afford to throw out all tenants who refuse to pay rent in a cohesive rent strike, or schools to dismiss all students who disobey an obnoxious school regulation if the students are united in their opposition.

Related to the tactic of clubbing the haves with their own book of rules and regulation is the tactic of goading them into errors such as violating their own rules or regulations. If they can be provoked into an obvious disruption of their own stated principles, then segments of the high-power group may become disaffected, with the resultant weakening of the haves. In addition, previously neutral third parties may, in response to the violations by those in power, swing their sympathies and support to the have-nots.

In general, it is a mistake to think that a high-power group is completely unified. Most groups have internal divisions and conflict among their most active members; further, only a small proportion of their members are likely to be active supporters of current policy. The conflicts among those who are active in the high-power groups and the distinction between active and passive members provide important points of leverage for the have-nots. The passive compliance of the inactive majority of the haves may disappear as their leaders are provoked into intemperate errors and as they are subject to ridicule and embarrassment by their inability to cope effectively with the persisting harassments and nuisances caused by the have-nots.

The power of the haves, as is true of any group, depends upon such tangibles as control over the instruments of force, an effective communication system, and an effective transportation system, and upon such intangibles as prestige and an aura of invincibility. While a low-power group may not be able to interfere seriously with the tangible bases of power of the haves without engaging in illegal, destructive actions of sabotage, it has many legal means of tarnishing and weakening their intangible sources of power. Ridicule and techniques of embarrassment are most effective weapons for this purpose. Here, as elsewhere, inventiveness and imagination play important roles in devising effective tactics.

Tactics of embarrassment and ridicule include the picketing of such people as slum landlords, key stockholders, management personnel of recalcitrant firms, and other such wielders of power in situations that are embarrassing to them—e.g. at their homes, at their churches, synagogues, or mosques, or at their social clubs. The advantage of such tactics as ridicule and embarrassment is that they are often enjoyable for those in low power and very difficult for those in high power to cope with without further loss of face.

4.7.4 Reducing the Power of the Oppressor

There are three strategies that are used to weaken oppressors: *divide and conquer*, *violence*, and *non-violence*. In prior sections, I have alluded to the *divide and conquer* strategy, and my emphasis there was on the recognition that there are often potential allies for the oppressed to be found among the oppressors. Even apart from recruiting allies among the oppressors, there is always the possibility of exploiting or creating divisions within this group. Various techniques can be employed in an attempt to create or increase the antagonism among different factions within the oppressors—e.g., planting rumors, creating incidents, making 'offers' that favor one faction over another, and distorting their processes of communication to one another in such a way that mistrust and hostility are fostered among the different factions.

4.7.4.1 Violence

As a strategy, violence has some positive features but, in my view, it has considerably greater negatives. Its positives are that it gets the attention of those in high power who have previously paid little attention to the oppressed and their needs. Additionally, it may be cathartic and psychologically empowering for those in low-power groups who feel enraged and humiliated by their oppression. Also, if well focused and executed, it may weaken the oppressed group.

Nelson Mandela, at one point, became convinced that nonviolent strategies were not being effective against the apartheid South African government, so he advocated that the African National Congress create a separate, secret group (MK), which would engage in violence. In planning the direction and form this group would take, Mandela (1995: 282–283) indicated that:

> We considered four types of violent activities: sabotage, guerilla warfare, terrorism, and open revolution. For a small and fledgling army, open revolution was inconceivable. Terrorism inevitably reflected poorly on those who used it, undermining any public support it might otherwise garner. Guerilla warfare was a possibility, but since the ANC had been reluctant to embrace violence at all, it made sense to start with the form of violence that inflicted the least harm against individuals: sabotage.
>
> Because it did not involve loss of life, it offered the best hope for reconciliation among the races afterward. We did not want to start a blood feud between white and black. Animosity between Afrikaner and Englishman was still sharp 50 years after the Anglo-Boer War; what race relations would be like between white and black if we provoked a civil war? Sabotage had the added virtue of requiring the least manpower.
>
> Our strategy was to make selective forays against military installations, power plants, telephone lines, and transportation links, targets that would not only hamper the military effectiveness of the state, but frighten National Party supporters, scare away foreign capital, and weaken the economy. This we hoped would bring the government to the bargaining table. Strict instructions were given to members of MK that we would countenance no loss of life.

Mandela was undoubtedly wise in advocating that the violence not be directed at people but rather directed at targets that would impair the government's political, economic, and military capabilities. Violence against people is apt to weaken the support of existing and potential allies, unify the oppressors, and lead to a vicious spiral of increasing irrational violence. The violence is irrational in that it is impelled by a thirst for vengeance rather than by an attempt to achieve strategic objectives. Violence of any sort against a powerful oppressor usually leads to an intensification of oppression rather than an increased readiness to engage in constructive negotiation.

The rare exceptions are when the violence by the oppressed is perceived by both the oppressed and the oppressor to be part of a rational appeal to the self-interest of the oppressor (i.e., an inevitable cost of refusing to engage in constructive negotiations for a change in the status quo which could be mutually beneficial). When the oppressor's response to violence is disproportionate, it may have the effect of delegitimizing the oppressor in the eyes of observers as well as in those of the oppressed. If the observers become active allies of the oppressed as a consequence of the oppressor's disproportionate reactive violence, then the balance of power may shift away from the oppressor to the oppressed.

I conclude that the use of violence by the oppressed against a much stronger oppressor is most likely to worsen its circumstances and, even in the unlikely possibility of a victory over the oppressor, it is apt to produce leadership among the former oppressed that is undemocratic and predisposed to employing violence in its leadership style.

4.7.4.2 Nonviolence

As a strategy, nonviolence is based on the premise that if we get what we want through violence, we will have created "a certain amount of harm, pain, injury, death, or destruction… We may in addition have created a climate of fear, distrust, or hatred on the part of those against whom we have used the violence. We may also have contributed to the transformation of ourselves into an insensitive or even cruel persons… Revolutions, even when they overcome violent resistance… often end up building the same sorts of abuses their promoters hoped to eliminate, just as wars set the stage for new wars" (Holmes 1990: 5).

In other words, the nonviolence strategy basically seeks to avoid the harmful effects of physical or psychological violence. Most approaches to nonviolence also assume that, in conflict, one should respect one's adversary and that even one's enemy is entitled to care and justice, to compassion and goodwill.

Sharp (1971), the most influential student of nonviolence, has identified at least 197 methods of nonviolent actions, which he groups into three categories:

(1) *Nonviolent protests* include marches, picketing, vigils, putting up posters, public meetings, and issuing and distributing protest literature. These methods are meant to produce an awareness of dissent and opposition to unjust policies and practices. Their impact can be large if they awaken the sense of injustice in influential potential allies who were not aware of the injustices being experienced.
(2) *Nonviolent noncooperation* includes refusal to comply with unfair rules, regulations, or orders, social or economic boycotts, boycotts of elections, general strikes, strikes, go-slow actions, rigid enforcement of rules, political jujitsu, civil disobedience, and mutiny. These methods are meant to disrupt the normal efficiency and functioning of the system controlled by the oppressor to indicate that the oppressed will no longer cooperate in their oppression.
(3) *Nonviolent interventions* include sit-ins, nonviolent obstructions of communication facilities, traffic, banks, public toilet facilities, etc., nonviolent invasions and occupancy, and creation of a parallel government. These methods are most coercive and disruptive to the functioning of the system, and they are most apt to produce a violent counter-response from those in power.

The use of nonviolent methods requires considerable self-discipline and courage. Systematic training of neophytes in the use of such methods by experienced practitioners makes their implementation more skillful and less dangerous. Training often involves role-playing and rehearsal of the appropriate actions to take in some of the typically difficult and dangerous situations that the non-violent participants may face as they engage in marches, refusals to comply with regulations, strikes, sit-ins, obstruction of traffic, or other nonviolent methods.

There has been no systematic research of which I am aware that attempts to determine the conditions under which nonviolent methods are likely to succeed or fail. There have been many instances of success as well as failure, and it is an area ripe for study (see Powers et al. 1997, for many case studies of nonviolent action). Based upon my very limited knowledge of these instances, I would hypothesize that nonviolent actions are most effective, when they are contesting clear and gross injustices, when they are well-publicized, when they are successful in recruiting others who are oppressed as well as allies among those who are not, and when they occur in a state that is reluctant to employ overwhelming force to repress the non-violent actions. In a state that controls the media and is repressive, success is unlikely unless the nonviolent actors are able to recruit the employees of the media and members of the police and armed forces to their side. In other words, nonviolent actors are likely to be most successful in democratic societies where repressive force against them is likely to be relatively moderate, and where they are apt to receive widespread, unfavorable publicity and to recruit allies to their cause. Thus, in the United States, the nonviolent civil rights movement was successful partly due to the widespread revulsion against the well-publicized violence used against them by public officials in the South. However, even in autocratically controlled states—such as apartheid South Africa, the Marcos government in the Philippines, the Shah's government in Iran, and the Milosevic government in Serbia, nonviolence was successful in overthrowing the governments. They were able to enlist the media and members of the armed forces to be against the repression of those seeking to change their oppressive, corrupt government.

Throughout much of the preceding discussion, I have emphasized the importance for low-power groups to use strategies and tactics which would develop allies among the high-power groups, among other low-power groups, and among third parties. Through their actions and resources, allies can play a vital role in not only awakening the sense of injustice in the oppressors but also increasing the bargaining power of the oppressed. Additionally, they can often facilitate a constructive, nonviolent process of conflict resolution and social change through the procedures and resources they make available in order to foster and maintain such a process.

4.8 Concluding Remarks

I conclude by stating that my objective in this article was to provide a generalized framework for characterizing oppression and the forms it takes, as well as to consider what keeps it in place and how it can be overcome. I hope that this framework can be usefully applied to understand and change oppressive relations between specific groups such as those between men and women, the rich and the poor, managers and workers, parents and children, and between different racial, religious, and ethnic groups.

References

Alinsky, S.D., 1971: *Rules for Radicals: A Practical Primer for Realistic Radicals* (New York: Random House).

Chaiken, S.L.; Gruenfeld, D.H.; Judd, C.M., 2000: in: Deutsch, M.; Coleman, P.T. (Eds.): *The Handbook of Conflict Resolution: Theory and Practice* (San Francisco: Jossey-Bass).

Crosby, F., 1982: *Relative Deprivation and Working Women* (New York: Oxford University Press).

de Tocqueville, A., 1947: *L'Ancien Regime* (Trans. Patterson, M.W.) (Oxford: Basil Blackwell).

Davies, J.C., 1962: "Toward a Theory of Revolution", in: *American Sociological Review*, 27: 5–19.

Deutsch, M., 1973: *The Resolution of Conflict: Constructive and Destructive Processes* (New Haven, CT: Yale University Press).

Deutsch, M., 1985: *Distributive Justice: A Social-Psychological Perspective* (New Haven: Yale University Press).

Deutsch, M., 1990: "Forms of Social Organization: Psychological Consequences", in: Himmelweit, H.T.; Gaskell, G. (Eds.): *Societal Psychology* (London: Sage).

Deutsch, M.; Coleman, P.T., 2000: *The Handbook of Conflict Resolution: Theory and Practice* (San Francisco: Jossey-Bass).

Deutsch, M.; Collins, M.E., 1951: *Interracial Housing* (Minneapolis, MA: University of Minnesota).

Deutsch, M.; Steil, J.M., 1988: "Awakening the Sense of Injustice", in: *Social Justice Research*, 2: 3–23.

Edelman, M., 1971: *Politics as Symbolic Action: Mass Arousal and Acquiescence* (Chicago: Markham).

Festinger, L., 1954: "A Theory of Social Comparison Processes", in: *American Psychologist*, 16: 1–11.

Freud, A., 1937.

Galtung, J., 1964: "Foreign Policy Position as a Function of Social Position", in: *Journal of Peace Research*, 1: 206–231.

Gil, D.G., 1998: *Confronting Injustice and Oppression* (New York, NY: Columbia University Press).

Gurr, T.R., 1970: *Why Men Rebel* (Princeton, NJ: Princeton University Press).

Harvey, J., 1999: *Civilized Oppression* (Lanham, MD: Rowman and Littlefield Publishers Inc.).

Hess, D.; Bonata, B. et al., 2005: *Encyclopedia of Selected Peaceful Societies* (http://www.peacefulsocieties.org).

Himmelstrand, V., 1969: "Tribalism, Nationalism, Rank-Equilibration, and Social Structures", in: *Journal of Peace Research*, 2: 81–104.

Holmes, R.L. (Ed.), 1990: *Nonviolence in Theory and Practice* (Belmont, CA: Wadsworth Publishing Co.).

Howell, S.; Willis, R. (Eds.), 1989: *Societies at Peace: Anthropological Perspectives* (New York: Routledge).

Kahneman, D.; Diener, E.; Schwarz, N. (Eds.), 1999: *Well-Beings: The Foundations of Hedonic Psychology* (New York: Russell Sage Foundation).

Kemp, G.; Fry, D.P. (Eds.), 2004: *Keeping the Peace: Conflict Resolution and Peaceful Societies around the World* (New York: Routledge).

Lane, R.E., 2000: *Loss of Happiness in Market Democracies* (New Haven: Yale University Press).

Lichtenberg, P., 1990: *Community and Confluence* (Cleveland, OH: Gestalt Institute of Cleveland Press).

Lind, E.A.; Tyler, T.R., 1988: *The Social Psychology of Procedural Justice* (New York: Plenum Press).
Mandela, N., 1994: *A Long Walk to Freedom: The Autobiography of Nelson Mandela* (London: Little Brown).
Mandela, N., 1995: *A Long Walk to Freedom: The Autobiography of Nelson Mandela*, Paperback edition (London: Little Brown).
Mead, M. (Ed.), 1937: *Cooperation and Competition: Among Primitive Peoples* (Boston: Beacon Press).
Neckerman, K.M. (Ed.), 2004: *Social Inequality* (New York: Russell Sage foundation).
Noel, L., 1994: *Intolerance: A General Survey* (Montreal: McGill-Queen's University Press).
Opotow, S.V., 1987: "Limits of Fairness: An Experimental Examination of Antecedents of the Scope of Justice" (Dissertation Information Service, University Microfilms International, Ann Arbor, MI. Order #8724072) [Dissertation Abstracts International, 48(08B), 2500].
Opotow, S.V., 1990: "Deterring Moral Exclusion: A summary", in: *Journal of Social Issues*, 46: 173–182.
Opotow, S.V., 1995: "Drawing the Line: Social Categorization, Moral Exclusion, and the Scope of Justice", in: Bunker, B.B.; Rubin, J.Z. (Eds.): *Conflict, Cooperation, and Justice* (San Francisco: Jossey-Bass): 347–369.
Opotow, S.V., 1996a: "Is Justice Finite? The Case of Environmental Inclusion", in: Montada, L.; Lerner, M. (Eds.): *Social Justice in Human Relations: Current Societal Concerns about Justice*, Volume 3 (New York: Plenum Press): 213–230.
Opotow, S.V., 1996b: "Affirmative Action, Fairness, and the Scope of Justice", in: *Journal of Social Issues*, 52: 19–24.
Opotow, S.V., 2001: "Social Injustice", in: Christie, D.J.; Wagner, R.V.; Winter, D.D. (Eds.): *Peace, Conflict and Violence: Peace Psychology for the 21st Century* (New York: Prentice-Hall): 102–109.
Perrucci, R.; Wysong, E., 1999: *The New Class Society* (Oxford: Rowman & Littlefield Publishers Inc.).
Pettigrew, T., 1967: "Social Evaluation Theory", in: Levine, D. (Ed.): *Nebraska Symposium on Motivation* (Lincoln: University of Nebraska Press).
Petty, R.E.; Wegener, D.T., 1998: "Attitude Change: Multiple Roles for Persuasion Variables", in: Gilbert, D.T.; Fiske, S.T.; Lindzey, G. (Eds.): *The Handbook of Social Psychology*, Volume 1, 4th edition (New York: McGraw-Hill).
Powers, R.S.; Vogele, W.B.; Krueglar, C.; McCarthy, R.M. (Eds.), 1997: *Protest, Power and Change: An Encyclopedia of Nonviolent Action from Act-up to Women's Suffrage* (New York, NY: Garland Publishing Inc.).
Runciman, W.C., 1966: *Relative Deprivation and Social Justice* (Berkeley: University of California Press).
Scholz, J.K.; Levine, K., 2004: "U.S. Black-White Inequality", in: Neckerman, K.M. (Ed.): *Social Inequality* (New York: Russell Sage Foundation).
Sharp, G., 1971: *The Politics of Nonviolent Action: An Encyclopedia of Thought and Action* (Philadelphia, PA: Pilgrim Press).
Sidanius, J.; Pratto, F., 1999: [No title or other information here].
Tajfel, H., 1982: *Social Identity and Intergroup Relations* (New York: Cambridge University Press).
Tyler, T.R.; Boeckman, R.J.; Smith, H.J.; Huo, Y.J., 1997: *Social Justice in a Diverse Society* (Boulder, Colorado: Westview Press).
Young, M.I., 1990: *Justice and the Politics of Difference* (Princeton, NJ: Princeton University Press).

Chapter 5
Interdependence and Psychological Orientation

5.1 Introduction[1]

In this chapter,[2] I shall examine the relations between types of psychological interdependence and psychological orientations. I shall employ the term *psychological orientation* to refer to a more or less consistent complex of cognitive, motivational, and moral orientations to a given situation that serves to guide one's behavior and responses in that situation. In brief, my theoretical analysis posits that distinctive psychological orientations are associated with the distinctive types of interdependence. I also assume that the causal arrow connecting psychological orientations and types of interdependence is bidirectional: A psychological orientation can induce or be induced by a given type of interdependence. Implicit in this view is the further assumption that each person has the capability to utilize the various psychological orientations and their associated cognitive, motivational, and moral orientations. Although individuals may differ in their readiness and ability to use the different orientations as a result of their cultural backgrounds, their personal histories, and their genetic endowments, people participate in diverse social relations in complex societies and these varied social relations require, and hence induce, different psychological orientations. Thus, my basic assumption is an evolutionary one: Namely, to cope with the psychological requirements of assorted types of social relations, people have developed the capacity to utilize psychological orientations as they are necessary in different situations.

This chapter is structured into four sections: (a) a discussion of types of interdependence, (b) a characterization of psychological orientations, (c) a discussion of

---

[1] This text was first published in 1982 as: Interdependence and psychological orientation. In V. Derlega and J.L. Grzelek (eds.), *Cooperation and Helping Behavior: Theories and Research* (New York: Academic Press, 1982). 15–42. Permission to republish this text was granted within Elsevier's global author re-use policies.

[2] The writing of this paper has been supported, in part, by a National Science Foundation Grant, BNS 77-16017.

P.T. Coleman and M. Deutsch, *Morton Deutsch: A Pioneer in Developing Peace Psychology*, SpringerBriefs on Pioneers in Science and Practice 30,
DOI 10.1007/978-3-319-15440-4_5

the relationship between types of interdependence and psychological orientations, and (d) a brief consideration of some relevant research. At the outset, I give notice to the reader that my chapter is not so ambitious as it may appear. I shall not attempt to discuss the full range of types of interdependence or psychological orientations. My aim is the more modest one of illustrating the potential fruitfulness of an idea that is still in the process of being formed, in the hope that doing so will stimulate other investigators to contribute to its development.

5.2 Types of Interdependence

Several years ago, 1 collaborated with Wish/Kaplan (1976) in research that sought to identify the fundamental dimensions of interpersonal relations. Based on this research, as well as earlier research by Triandis (1972) and Marwell/Hage (1970) and later research by Wish/Kaplan (1977), it seems reasonable to assert that the fundamental dimensions of interpersonal relations include the following:

1. *Cooperation-competition.* This dimension is referred to variously in the social psychological literature. I have characterized it as *promotive versus contrient interdependence* (Deutsch 1949a) or as a *pro-con* dimension (Deutsch 1962). Triandis (1972) refer to it as *association-disassociation*, Kelley/Thibaut (1978) use the term *correspondence-noncorrespondence,* and it has been labeled *love-hate, evaluative, positive-negative interpersonal disposition, friendly-hostile,* etc. by other investigators. In the Wish et al. (1976) study, scales of the following sort were strongly weighted on this dimension: "Always harmonious versus always clashing," "very co-operative versus very competitive," "very friendly versus very hostile," "have compatible versus incompatible goals and desires," "very productive versus very destructive," "find it easy versus difficult to resolve conflicts with each other," "very altruistic versus very selfish," "very fair versus very unfair."[3]
   Such interpersonal relations as "close friends," 'teammates,' and 'coworkers' are at the cooperative end of the dimension, whereas "political opponents," "personal enemies," "divorced couple," and "guard and prisoner" are toward the competitive end. The social psychological processes and consequences associated with this dimension have been extensively investigated in my theorizing and research (Deutsch 1949a, b, 1962, 1973).
2. *Power distribution ('equal' vs 'unequal').* This dimension has been given various labels: Triandis (1972) characterizes it as *superordination-subordination,* Kelley (1979) describes it in terms of *mutuality of interdependence,* and others have used such terms as *dominance-submission, potency,* and *autonomy-control.* Scales such as the following are strongly weighted on this dimension: "exactly

[3] Although this research studied the *perceptions* of interpersonal relations, I see no reason to doubt that the identified dimensions are fundamental aspects of interpersonal relations.

equal versus extremely unequal power," "very similar versus very different roles and behaviors," and "very democratic versus very autocratic attitudes." "Business partners," "close friends," and "business rivals" are at the 'equal' end; "master and servant," "teacher and pupil," "parent and child," and "guard and prisoner" are at the 'unequal' end. The social psychological processes and consequences associated with this dimension are reviewed in Cartwright/Zander (1968).

3. *Task-oriented versus social-emotional.* This dimension has been labeled *intimacy* by Triandis (1972) and Marwell/Hage (1970) and *personal* by Kelley (1979). Others have identified it as *personal-impersonal, subjective versus objective, particularistic versus universalistic,* or *emotionally involved versus emotionally detached.* The two following scales are strongly weighted on this dimension: "pleasure-oriented versus work-oriented," and "emotional versus intellectual." Such interpersonal relations as "close friends," "husband and wife," and 'siblings' are at the social-emotional end of the dimension; "interviewer and job applicant," "opposing negotiators," "supervisors and employees," and "business rivals" are at the task-oriented end. Bales's (1958) distinction between social-emotional and task-oriented leaders of groups is relevant; the former focuses on the solidarity relations among group members, and the latter focuses on the external task and problem-solving activities of the group. Earlier, I made a similar distinction between *task junctions* and *group maintenance functions* (Deutsch 1949a, b), which was elaborated in a paper by Benne/Sheats (1948). The sociological distinction between *gemeinschaft* and *gesellschaft* groups also reflects this basic dimension of social relations.
4. *Formal versus informal.* Wish/Kaplan (1977) have shown that this dimension can be separated from the preceding one. It appears to be the same as the dimension of *regulation* identified by Marwell/Hage (1970).
   In an informal relationship, the definition of the activities, times, and locations involved in the relationship are left largely to the participants; in a formal or regulated relationship, social rules and norms largely determine the interactions among those involved. Such scales as "very formal versus very informal" and "very flexible versus very rigid" reflect this dimension. Relations within a bureaucracy tend to be formal, whereas relations within a social club tend to be informal; also, relations between equals are more likely to be informal than relations between unequals. Formal, bureaucratic relationships have been the subject of extensive discussions by such socio-logical theorists as Weber (1957) and Merton (1957).
5. *Intensity or importance.* This dimension has to do with the intensity or superficiality of the relationship. Kelley (1979) suggests that it reflects the degree of interdependence (or dependence) in the relationship. Such scales as the following are strongly weighted on it: "very active versus very inactive," "have intense versus superficial interactions with each other," "have intense versus superficial feelings toward each other," and "important versus unimportant to the individuals involved." "Casual acquaintances," "second cousins," and

"salesman and customer" are at the superficial end of this dimension; "parent and child," "husband and wife," "psychotherapist and patient" are at the intense end.

Several other dimensions of interpersonal relations have been identified, including the enduring or temporary nature of the relationship, its voluntary or involuntary character, its public versus private nature, its licit or illicit quality, and the number of people involved in the relationship. It is beyond the scope of this paper to consider these other dimensions.

Table 5.1 presents the first four dimensions in dichotomous form and provides illustrations of the types of interpersonal relations and types of interpersonal activities that could occur in each of the 16 regions of this four-dimensional space. (I have selected illustrations from the "more intense" rather than "less intense" end of the intensity dimension.) It is, of course, an oversimplification to dichotomize each of the dimensions, but it is a reasonable place to start. If the reader were to blank out the illustrations in Table 5.1 and attempt to provide other examples, he or she would probably discover that the dimensions are correlated. It is easier to find illustrations for some of the 16 regions than others; some of the regions are undoubtedly more heavily populated than others.

Thus, social-emotional relations or activities are more likely to be informal than the task-oriented ones, especially if there are relatively more people involved in the task-oriented ones. Also, there appears to be a positive linkage between the informality of the relation or activity and its equality, so that it is more difficult to find unequal, informal relations and activities than equal, informal ones. Moreover, there is evidently a positive association between the cooperativeness and informality of a relation or activity. Similarly, there appears to be a positive connection between the equality of an activity or relation and its cooperativeness. Additionally, there is likely to be a positive association between the social-emotional nature of a relation or activity and its cooperativeness. Further, one can expect that social-emotional relations and activities will more frequently be intense than task-oriented ones. Also, interpersonal relations or activities that are extremely cooperative or competitive, rather than moderately so, will be more intense.

The foregoing hypothesized correlations among the dimensions suggest which regions of the interpersonal space will be heavily populated and which will not be.[4] (See Wish/Kaplan 1977, for some support for the hypothesized correlations.) Thus, one would expect more interpersonal relations and activities (particularly, if they are stable and enduring) to be clustered in the cooperative, equal, informal, and social-emotional region (Cell 1 in Table 1.1), which I shall label the *intimacy* region, than in the competitive, equal, informal, and social-emotional region (Cell 3), which I shall label the *antagonistic* region. Intense competitive relations or activities are more likely to be stable and enduring if they are regulated or formal rather than unregulated. Thus, one would expect Cell 7 ('rivalry') to be more populated than Cell 3

---

[4] *INDSCAL*, the multidimensional scale analysis procedure used in the Wish et al. (1976) and the Wish/Kaplan (1977) studies does not force the identified dimensions to be orthogonal.

Table 5.1 Sixteen types of social relations

	Social-emotional		Task-oriented	
	Informal	Formal	Informal	Formal
Cooperative				
Equal	1	5	9	13
	Intimate	Fraternal	Problem-solving	Organized cooperation
	Lovers	Club members	Colleagues	Task force members
	Love-making	Social party	Staff meeting	Working together with differentiated responsibilities to solve problems
Unequal	2	6	10	14
	Caring	Protecting	Educational	Hierarchical organization
	Mother-child	Police officer-child	Professor-graduate student	Supervisor-employee
	Nursing	Helping	Working together informally on research project under professor's direction	Supervisor assigning employee to do certain task
Competitive				
Equal	3	7	11	15
	Antagonistic	Rivalrous	Competitive	Regulated competition
	Personal enemies	Divorced couple	Contestants in informal game	Business rivals
	Fighting	Custody suit	Trying to score points against the other	Bidding against one another for a contract
Unequal	4	8	12	16
	Sadomasochistic	Dominating	Power struggle	Regulated power struggle
	Bully-victim	Expert-novice	Authority-rebel	Guard-prisoner
	Tormenting	Intimidating	Guerilla warfare	Ordering prisoner to keep in step

Note Each cell characterizes a type of social relation by labeling the relation (first entry), naming people who might be in such a relationship (second entry), and describing an activity that might occur in such a relationship (third entry)

('antagonism'); similarly, Cell 8 ("sado-masochism") and for Cells 15 ("regulated competition") and 16 ("regulated power struggle") would be more populated compared to their respective unregulated cells.

Intense, cooperative, task-oriented relations or activities are more apt to be equal and informal than otherwise, unless there are clear status differences among the people involved (i.e., to be located in Cell 9 rather than in Cell 10, 13, or 14). However, the demands of large-scale cooperative tasks involving more than small numbers of people are apt to require a formal, hierarchical (i.e., unequal) organization for the tasks to be worked on effectively and efficiently. Thus, one could expect many hierarchically organized cooperative relations and activities to be found in Cell 14 ("hierarchical organization"). Yet the nature of such unequal relations as superordinate-subordinate ones in organizations, especially when they are not strongly legitimated for those in the subordinate position, is such as to produce conflict over the power differences. Hence, this type of relation is rarely free of strong competitive elements. It follows, then, that some superordinate-subordinate relations in hierarchically organized systems will have the character of power struggles, and these would be more appropriately classified as belonging to Cell 16.

5.3 Psychological Orientations

In writing an earlier draft of this chapter, I entitled this section "Modes of Thought." This earlier title did not seem to be a sufficiently inclusive label. It appeared to me evident that cognitive processes differ in different types of social relations, and I wanted to sketch out the nature of some of these differences. However, I also thought that the psychological differences among the different types of social relations were not confined to the cognitive processes. Different motivational and moral predispositions were also involved. It has been customary to consider these latter predispositions as more enduring characteristics of the individual and to label them "personality traits" or "character orientations." Since my emphasis is on the situationally induced nature and, hence, temporariness of such predispositions, such labels also did not seem fitting for the material in this section. I have used the term *psychological orientation* to capture the basic theme of this section: People orient themselves differently to different types of social relations, and different orientations reflect and are reflected in different cognitive processes, motivational tendencies, and moral dispositions.

5.3.1 The Cyclical Relation Between Psychological Orientations and Social Relations

Figure 5.1 depicts in schematic form my view of this association between psychological orientations and social relationships, as well as some other factors influencing both of them. It was stimulated by Neisser's (1976) conception of the perceptual cycle but is a radical modification of it. My emphasis, like Neisser's, is on the cyclical and active process involved in the connection among the elements.

In characterizing this cyclical, active process, one can start at any point in the cycle. In practice, where one starts will usually be determined by what one manipulates as one's independent variable. The non-manipulated variables will be considered to be the dependent ones.

Let us suppose, for example, that as an experimenter, I lead a subject to have the psychological orientation typical of a mutually promotive, interdependent relationship toward another. This, in turn, will lead the subject to have some characteristic interactions with the other and these, in turn, will have some effects upon both the subject and the other that will provide evidence as to the type of relationship that exists between them. Finally, this will validate as appropriate, or invalidate as inappropriate, the subject's psychological orientation and require its modification. Here the cycle is A–Or, I might begin at a different point, C, inducing the subject to believe that he was in a mutually promotive, interdependent relationship. This would, in turn, lead the subject to have a psychological orientation toward the relationship that has specifiable characteristics and this, in turn, would lead to certain interactions with the other, etc. Here the cycle is C−^A—>B−H>C. Or, I might commence by leading the subject to interact with the other in specified ways that, in turn, would produce certain consequences that, in turn, would produce evidence as to what kind of relationship the subject was in, etc. Here the starting point is B and the cycle goes on to C, etc.

Several other features of Fig. 5.1 merit comment. I assume that the two parts of each triangle can affect one another:

1. One's psychological orientation to one's present social relationship can be affected by and can also affect one's desires with regard to that relationship. Thus, if one has a desire for a cooperative relationship but a contrient orientation to the other, one may change either one's desire or one's orientation, depending on which is less strongly rooted.
2. One's present social relations with another can influence or be influenced by the potential one sees for the development of the relationship. If I experience the present relationship as a destructive one, I might not see it as having a future; on the other hand, if I see the potential of developing a warm, loving relationship, I might be more positive toward an initially difficult relationship than I might otherwise be.
3. The nature of one's actions and reactions in a relationship can affect as well as be affected by the normative definitions that exist regarding interactions in a given social relationship. Although culturally determined normative definitions often govern the meanings of social interactions early in a relationship, relations tend to build up their own idiosyncratic normative definitions as a result of repeated interactions that may be peculiar to the particular relationship.

There is, of course, a tendency for the two parts of each triangle to be consistent with one another. When they are not, one can expect a more complex psychological structure than the one depicted in Fig. 5.1. For example, if the present and future characteristics of the social relationship are perceived to be inconsistent with one another, the time perspective dimension of the relationship will be very prominent.

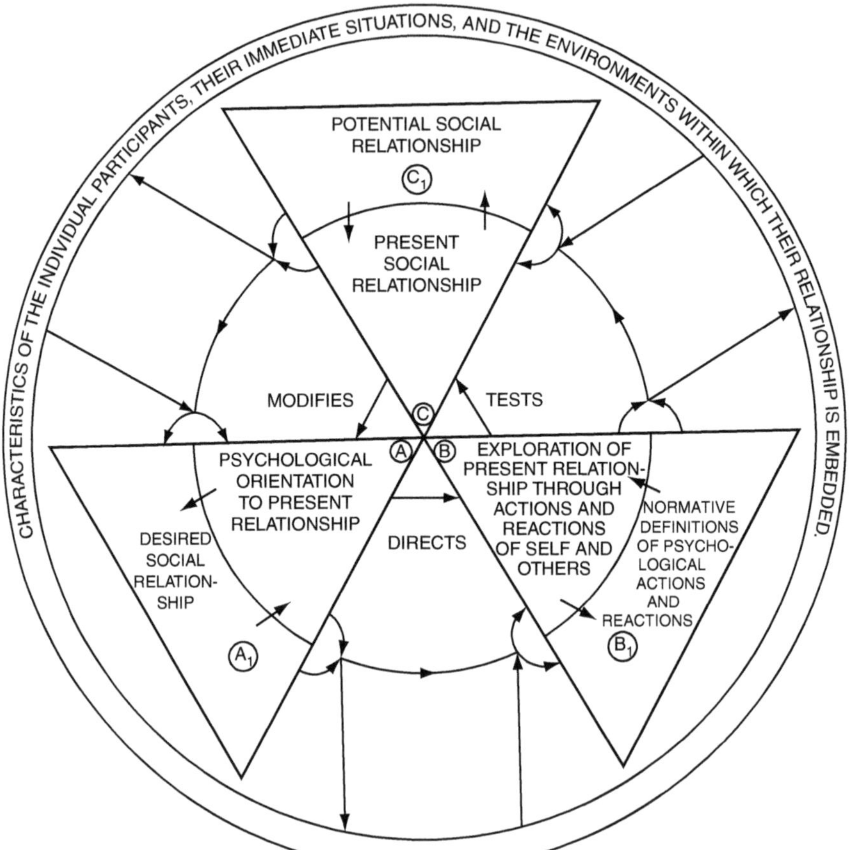

Fig. 5.1 The circular relation between an individual's psychological orientation and the type of social relationship in which he or she is involved

If there is an inconsistency between the desired social relationship and the present psychological orientation, the reality dimension will be very prominent. It is beyond the scope of the present paper to consider these important psychological aspects of social relations; Lewin (1951) makes suggestive remarks about these dimensions of the life space in his writings.

Surrounding the triangles of Fig. 5.1 is the 'objective' world of the participants; this includes the characteristics of the individual participants, their immediate situations, and the environment within which their relationship is embedded. I have characterized this objective world as sending causal arrows to all of the elements involved in the psychological orientation-social relations cycle and also as receiving causal arrows from these elements. The nature of the participants and their immediate situations, as well as their environment, affect their social relations, their psychological orientation, and their interactions, and these phenomena, in turn, affect the participants and the realities confronting them. In this larger cycle, it is the variables that one considers independent that one manipulates.

5.3.2 The Nature of Psychological Orientations

5.3.2.1 Cognitive Orientations

In recent years, scholars in a number of different disciplines—cognitive psychology, social psychology, sociology, linguistics, anthropology, and artificial intelligence—have utilized such terms as *schema, script,* and *frame* to refer to the *structures of expectations* that help orient the individual cognitively to the situation confronting him. I shall employ the term *cognitive orientation* as being essentially the same as these terms. In the view being presented here, the person's cognitive orientation to his situation is only one aspect of his psychological orientation to a social relationship. Other aspects include his motivational orientation and his moral orientation.

The term *schema* goes back to Bartlett (1932) who, much influenced by the work of the neurologist Head (1920), emphasized the constructive and organized features of memory as opposed to the notion of memory as passive storage. The term *script* derives from the work of Abelson (1975, 1976) and Schanck/Abelson (1977), who also stress that people have organized knowledge of a stereotypic form about most recurrent situations they encounter. Abelson (1975) defines a script as a "coherent sequence of events expected by the individual, involving him either as a participant or as an observer [p. 33]." He goes on to postulate that "cognitively mediated social behavior depends on the joint occurrence of two processes: (a) the selection of a particular script to represent the given social situation and (b) the taking of a participant role within that script [pp. 42–43]." The term *frame* was introduced by Bateson (1955) to explain how individuals exchange signals that allow them to agree on the level of abstraction at which any message is intended; for example, whether the message is intended as serious or playful. Goffman (1974) has generalized Bateson's discussion of frames in an extended analysis of how individuals, as they attend to any current situation, face the question, "What is it that's going on here?"

Underlying the concepts of schema, script, and frame is the shared view that people approach their social world actively, with structured expectations about themselves and their social environments that reflect their organized beliefs about different social situations and different people. Our structured expectations make it possible for us to interpret and respond quickly to what is going on in specific situations. If our expectations lead us to in-appropriate interpretations and responses, then they are likely to be revised on the basis of our experiences in the situation. Or, if the circumstance confronting us is sufficiently malleable, our interpretations and responses to it may help to shape its form.

Schemas, scripts, or frames may be very concrete and specific—for example, how to work together with a particular person on a given task—or they may be rather abstract and general—for example, what is involved in a competitive as compared to a cooperative relation. In any society that provides a variety of situations in which different areas in the multidimensional space of social relations (the space being composed from the dimensions that were described in the first major section of this chapter) are well-represented, it is likely that rather abstract schemas or scripts will develop to characterize the types of relations depicted in Table 1.1. Such scripts, or cognitive orientations, are a central component of what I am here terming psychological orientations.

It is important for the participants in a particular social relationship to know "what's going on here"—to know the actors, the roles they are to perform, the relations among the different roles, the props and settings, the scenes, and the themes of the social interaction. However, everyday social relations are rarely as completely specified by well-articulated scripts as is social interaction in a play in the traditional theatre; ordinary social interactions have more the qualities of improvisational theatre, in which only the nature of the characters involved in the situation is well-specified and the characters are largely free to develop the details of the skeletonized script as they interact with one another.

The improvisational nature of most social relations—the fact that given types of social relations occur in widely different contexts and with many different kinds of actors—makes it likely that relatively abstract or generalized cognitive orientations, schemas, or scripts will develop for the different types of social relations. I assume that people are implicit social psychological theorists and, as a result of their experience, have developed cognitive schemas of the different types of social relations that, though usually not articulated, are similar to those articulated by theorists in social psychology and the other social sciences. Undoubtedly, at this early stage of the development of social science theory, the unarticulated conceptions of the average person are apt to be more sophisticated than the articulated ones of the social scientists.

5.3.2.2 Motivational Orientations

Just as different cognitive orientations are associated with the different types of social relations, so also are different motivational orientations. A motivational orientation toward a given social relationship orients one to the possibilities of gratification or frustration of certain types of needs in the given relationship. To the cognitive characterization of the relationship, the motivational orientation adds the personal, subjective features arising from one's situationally relevant motives or need-dispositions.

The motivational orientation gives rise to the cathexis of certain regions of the cognitive landscape, making them positively or negatively valent, and highlights the pathways to and from valent regions. It gives the cognitive map a dynamic character. It predisposes one to certain kinds of fantasies (or nightmares) and to certain kinds of emotions. It orients one to such questions as "What is to be valued in this relationship?" and "What do I want here and how do I get it?"

It is evident that different types of social relations offer different possibilities of need gratification. It would be unreasonable, for example, to expect one's need for affection to be gratified in a business transaction and inappropriate to expect one's financial needs to be fulfilled in an intimate relationship. In the third section of the chapter I shall attempt to characterize briefly the motivational orientations associated with the polar ends of the different dimensions of interpersonal relations.

5.3.2.3 Moral Orientations

A moral orientation toward a given social relationship orients one to the mutual obligations, rights, and entitlements of the people involved in the given relationship. It adds an "ought to," 'should,' or obligatory quality to a psychological orientation. The moral orientation implies that one experiences one's relationship not only from a personal perspective, but also from a social perspective that includes the perspective of the others in the relationship. A moral orientation makes the experience of injustice more than a personal experience. Not only is one personally affected, but so are the other participants in the relationship, because its value underpinnings are being undermined. The various participants in a relationship have the mutual obligation to respect and protect the framework of social norms that define what is to be considered as fair or unfair in the interactions and outcomes of the participants. One can expect that the moral orientation, and hence what is considered fair, will differ in the different types of social relations.

5.4 The Relationship Between Types of Interdependence and Psychological Orientations

In this section, I shall characterize the psychological orientations that are associated with the dimensions of cooperation-competition, power, task-oriented versus social-emotional, and formal versus informal. For each of the four dimensions depicted in Table 1.1, I shall describe the cognitive, motivational, and moral orientations that typify the dimension.

5.4.1 Cooperation-Competition

5.4.1.1 Cognitive Orientation

The cooperative-competitive dimension seems so fundamental to social life that one would assume a well-developed innate predisposition to developing abstract cognitive orientations to help an individual define quickly whether "what's going on here" is 'good' for him or 'bad' for him. With additional experience and further psychological differentiation and integration, the basic cognitive schema of cooperation-competition should emerge: We are 'for' one another or 'against' one another; we are linked together so that we both gain or lose together or we are linked together so that if one gains, the other loses. This basic schema has many implications (see Deutsch 1949a, b, 1962, 1973, 1979 for an elaboration of these implications). It leads an individual holding it to expect that in a cooperative relation, the other will be pleased by the individual's effective actions and ready to help him or her achieve success. The individual will expect the opposite to be true in a competitive relationship. If one believes one is in a cooperative relationship and the other is displeased by one's effective actions, one will wonder, "What is going on here?" "Am I in the kind of relation that I think I am in?" "What can I do to find out what is going on here?"

5.4.1.2 Motivational Orientation

In a cooperative relation, one is predisposed to cathect the other positively; to have a trusting and benevolent attitude toward the other; to be psychologically open to the other; to be giving as well as receptive to the other; to have a sense of responsibility toward the other and toward the mutual process of cooperation; to see the other as similar to oneself; etc. One is also predisposed to expect the other to have a similar orientation toward oneself. Murray's (1938: 175–177) description of the *need for affiliation* captures much of the essence of this motivational orientation. It is clear that the specific quality of this orientation will be very much influenced by what type of cooperative relation is involved: social-emotional or task-oriented, equal or unequal, formal or informal, intense or superficial.

In a competitive relation, one is predisposed to cathect the other negatively; to have a suspicious and hostile, exploitative attitude toward the other; to be psychologically closed to the other; to be aggressive and defensive toward the other; to seek advantage and superiority for the self and disadvantage and inferiority for the other; to see the other as opposed to oneself and basically different; etc. One is also predisposed to expect the other to have the same orientation. Murray's (1938) description of the *need for aggression* (pp. 159–161) and *need for defendance* (pp. 194–195) as well as the associated needs for 'infavoidance' (pp. 192–193) and 'counteraction' (pp. 195–197) seem to characterize many of the basic features of this motivational orientation. The specific quality of this motivational orientation will be determined by the type of competitive situation: task-oriented or social-emotional, equal or unequal, formal or informal, intense or superficial. In addition, it will be colored by one's conception of one's chances of winning or losing.

5.4.1.3 Moral Orientation

Although the specific character of the moral orientations associated with cooperation and competition will also depend on other features of the social relationship, it seems evident that cooperation and competition elicit different types of moral orientations. The moral orientation linked with cooperation is a tendency toward egalitarianism. This tendency underlies a general conception of justice that Rawls (1972) has expressed as follows: "All social values—liberty and opportunity, income and wealth, and the bases of self-respect—are to be distributed equally unless an unequal distribution of any, or all, of these values is to everyone's advantage [p. 62]." The moral orientation connected with cooperation fosters mutual respect and self-respect and favors equality as a guiding value to be breached only when inequality brings greater benefits and advantages to those less fortunate than they would otherwise have been if all were treated equally. Given this moral orientation, as Rawls (1972) points out, "Injustice, then, is simply inequalities that are not to the benefit of all [p. 62]."

In contrast, the moral orientation linked with competition sanctions in-equality and legitimates a win-lose struggle to determine who will have superior and who will have inferior outcomes in a competitive relationship. Depending on other features of the relationship, the struggle may be regulated so that the competition takes place under fair rules (as in a duel of honor) and one's moral orientation will include an obligation to obey the rules, or the struggle may be a "no-holds-barred" one in which any means to defeat the other can be employed. An active state of competition implies that the competitors do not mutually perceive and accept a superior-inferior relationship between them: If they do, and they continue to wage competition, then they are violating the moral imperatives of competitive justice. Thus, it is part of the moral orientation of competition for a victor to accept the defeat of someone who acknowledges being vanquished without continuing to beat the defeated one.

5.4.2 Power ('Equality' vs. 'Inequality')

5.4.2.1 Cognitive Orientation

The basic schema of "relationship power" (Deutsch 1973) has to do with the relative power of the participants in a relationship to benefit or harm or persuade one another and, hence, their relative power to influence one another. In a relationship of unequal power, it is expected that the more powerful member will be advantaged and the less powerful one will be disadvantaged whenever their interests are opposed. Hence, it is considered better to be in the more- rather than the less- powerful position in a competitive relationship. The competitive branch of the unequal power schema highlights the roles of 'victor' and 'vanquished'; the equal power schema orients more to continuing struggle. In both competitive branches, the use of tactics of coercion, intimidation, and power bluffs are made salient. Even in a situation where the more- and less-powerful members have congruent interests, the less-powerful member is expected to be more dependent on the other and, hence, more likely to engage in ingratiating behavior. The cooperative branch of the unequal power schema emphasizes the orientation toward responsibility in the high power position and of respectful compliance from the low power position; the equal power schema orients more toward mutual responsibility and respect. Both cooperative branches make salient the use of the more positive forms of power: persuasion rather than coercion, benefits rather than harms, legitimate rather than illegitimate power, etc.

5.4.2.2 Motivational Orientation

In an equality relation, one is predisposed to consider that the other is entitled to the same esteem and respect as oneself. The equality of power is likely to signify that the different participants in a relationship have the same value. Respect and esteem are more valuable if they are received from those whom one respects; equal status relations represent the optimum distribution for the mutual support of self-esteem. The need-dispositions related to self-esteem and self-respect seem to underlie this motivational orientation. The need for self-esteem involves the need to have a sense of the worthiness of one's goals and a sense of confidence in one's ability to fulfill one's intentions; the need for self-respect involves the need to have a sense of one's moral worth, of one's equal right to justice and fair treatment. It undermines one's sense of belonging to a moral community to be treated more fairly or less fairly than others and this, in turn, weakens the foundations of self-respect. Hence, one's self-respect is more firmly grounded in relationships where one can feel the others are also entitled to respect.

Similarly, the confidence in oneself that is connected with a secure self-esteem is fostered by association and comparison with people who are similar in status rather than with those who are higher or lower.

In an unequal relationship, one is predisposed either to take a more dominant or a more subordinate role or to resist the inequality. Murray (1938) has characterized the different aspects of the *need for dominance*. It is manifest in the desire to control, influence, direct, command, induce, dictate, supervise, instruct, or lead. In a competitive situation, the need for dominance will often be fused with the need for aggression and will lead to attempts to coerce and force the other to comply with one's desires. In a cooperative situation, it will often be fused with the need for nurturance and will lead to a protective, guiding, and caring orientation toward the other. Different needs are associated with the submissive role, depending on whether it occurs in a cooperative or competitive *context*. In a *cooperative* context, Murray's (1938: 154–156) description of the *need for deference* seems appropriate. It involves a readiness to follow, to comply, to emulate, to conform, to obey, to defer, to admire, to revere, to be suggestible, to heed advice, and otherwise to accept the superior authority of the other. In a competitive relationship, the need-disposition associated with the acceptance of the inferior role is well characterized in Murray's (1938: 161–164) description of the *need for abasement*. This disposition is reflected in the tendency to submit passively, to accept blame, to surrender, to seek punishment or pain, to be servile, to be resigned, to acquiesce, to be timorous, to give in, and to allow oneself to be bullied. It is evident that the subordinate role in an unequal relationship may be difficult to accept and may be resisted. The resistance to an unequal relationship will be evidenced in aspects of what Murray has termed the *need for autonomy* (pp. 156–159) and the *need for rejection* (pp. 177–180). The need for autonomy is characterized by the tendency to resist coercion and restraint, to be defiant and rebellious in relation to arbitrary authority, to be independent of social ties, and to be a nonconformist. The need for rejection is reflected in the tendency to separate oneself from a negatively cathected other; to reject a disliked superior other; to out-snub a snob; to exclude, abandon, expel or remain indifferent to an inferior other.

5.4.2.3 Moral Orientation

As the preceding discussion of motivational orientations would suggest, there are a number of different moral orientations connected with equality and inequality: Other features of the relationship, in addition to the distribution of power within it, will determine the nature of the moral orientation that will be elicited. Thus, in a cooperative, equal relationship, one would expect the kind of egalitarian relationship described in the section on the moral orientation associated with cooperation-competition. In a cooperative, unequal relationship, the moral orientation obligates the more powerful person to employ his power in such a way as to benefit the less powerful one, not merely himself. In such a relationship, the less powerful one has the obligation to show appreciation, to defer to, and honor the more powerful person. These obligations may be rather specific and limited if the relationship is task-oriented or they may be diffuse and general if the relationship is a social-emotional one.

In an equal, competitive relationship, one's moral orientation is toward the value of initial equality among the competitors and the subsequent striving to achieve superiority over the others. This orientation favors "equal opportunity" but not "equal outcomes": The competitors start the contest with equal chances to win, but some win and some lose. In an unequal, competitive relationship, the moral orientations of the strong and the weak support an exploitative relationship. The strong are likely to adopt the view that the rich and powerful are biologically and, hence, morally superior; they have achieved their superior positions as a result of natural selection; it would be against nature to interfere with the inequality and suffering of the poor and weak; and it is the manifest destiny of superior people to lead inferior peoples. The beatitude of those in powerful positions who exploit those in weaker positions appears to be, "Blessed are the strong, for they shall prey upon the weak" (Banton 1967: 48). In an unequal, competitive relationship, the weak are apt to *identify with the aggressor* (Freud 1937), to adopt the moral orientation of the more powerful, and to feel that their inferior outcomes are deserved. Or, they may feel victimized. If so, they may either develop a revolutionary moral orientation directed toward changing the nature of the existing relationship, or they may develop the moral orientation of being a victim. The latter orientation seeks to obtain secondary gratification *from* being morally superior *to the victimizer*: "It's better to be sinned against than to sin"; "the meek shall inherit the earth."

5.4.3 Task-Oriented Versus Social-Emotional

5.4.3.1 Cognitive Orientation

The basic schema here has to do with the focus of involvement. In a task-oriented relationship, one expects the attention and the activities of the participants to be directed toward something external to their relationship, whereas in a social-emotional relationship, one expects much of the involvement to be centered on the relationship and the specific persons in the relationship. This difference in focus leads one to expect a relationship that is primarily task-oriented to be impersonal in the sense that the actual accomplishment of the task is more important than the identity of the persons involved in accomplishing it and the nature of their personal relationships. In a task-oriented relationship, people who can perform equally well on the task are substitutable for one another. The personal identity and the unique individuality of the performer have little significance in such a relationship.

In contrast, in a social-emotional relationship, the personal qualities and identity of the individuals involved are of paramount importance. People are not readily substitutable for one another. Using Parsonian terminology, in a task-oriented relation, people are oriented to one another as *complexes of performances*—that is, in terms of what each *does;* in a social-emotional relationship people are oriented to each other as *complexes of qualities*—that is, in terms of what each *is*. Also, in a task-oriented relationship, one's orientation toward the other is universalistic—that is, one applies general standards that are independent of one's particular relationship with the other; in a social-emotional relationship, one's orientation is particularistic—that is, one's responses to the other are determined by the particular relatedness that exists between oneself and the other.

In a task-oriented relationship, one is oriented to making decisions about which means are most efficient in achieving given ends.[5] This orientation requires an abstract, analytic, quantifying, calculating, comparative mode of thought in which one is able to adopt an affectively neutral, external attitude toward different means in order to be able to make a precise appraisal of their comparative merit in achieving one's ends. One orients to other people as instrumental means and evaluates them in comparison or competition with other means. In *contrast,* in a social-emotional relationship, one is oriented to the attitudes, feelings, and psychological states of the other as ends. This orientation requires a more holistic, concrete, intuitive, qualitative, appreciative-aesthetic mode of thought in which

[5] 1 caution the reader not to conclude from this sentence or from anything else in this chapter that relationships that are exclusively task-oriented will be *more productive than*those that have a mixture of task-orientedness and social-emotional orientedness. Effective group functioning on tasks, for example, requires attention to "group maintenance" as well as to "task functions" (Deutsch 1949b).

one's own affective reactions help one to apprehend the other from the 'inside.' Other people are oriented to as unique persons rather than as instruments in which aspects of the person are useful for particular purposes.

5.4.3.2 Motivational Orientation

A task-oriented relationship tends to evoke achievement-oriented motivations. Achievement motivation has been discussed extensively by Murray (1938), McClelland et al. (1953), Atkinson/Feather (1966), and Weiner (1974). Here I wish merely to indicate that it consists not only of the egoistic motivations to achieve success and to avoid failure; motivations related to using one's capabilities in worthwhile activities may also be involved. Additionally, since achievement motivation is often instrumentally oriented to serve an adaptive function in relation to the external environment characterized by a scarcity rather than abundance of resources, it usually contains an element of motivation that is oriented toward rational, efficient accomplishment of the task. Further, since task- oriented relationships are primarily instrumental rather than consummatory in character, they require a motivational orientation that accepts delay-in gratification and that obtains satisfaction from disciplined activity oriented toward future gratification.

A number of different motivational orientations are likely to be elicited in social-emotional relationships: affiliation, affection, esteem, play, sentience, eroticism, and nurturance-succorance. The primary feature of these different need-dispositions as they are manifested in social-emotional relationships is that they are focused on the nature of the person-to-person (or person-to-group) relationship: They are oriented toward giving and receiving cathexes; toward the attitudes and emotions of the people involved in the relationship; toward the pleasures and frustrations arising from the interaction with the particular others in the given relationship. Although past experiences and future expectations may affect how one acts toward others and how one interprets the actions of others in a social-emotional relationship, such a relationship—if it is a genuine one—is not instrumental to other, future goals; it is an end in itself. In this sense, the need-dispositions in a social-emotional relationship are oriented toward current rather than delayed gratification.

5.4.3.3 Moral Orientation

The moral orientation in a task-oriented relationship is that of utilitarianism. Its root value is maximization: People should try to get the most out of situations. Good is viewed as essentially quantitative, as something that can be increased or decreased without limit (Diesing 1962: 35). A second element in this moral orientation is the means-end schema, in which efficient allocation of means to achieve alternative ends becomes a salient value. A third element is impartiality in the comparison of means, so that means can be compared on the basis of their merit in achieving given ends rather than on the basis of considerations irrelevant to the means-end relationship. In Parsonian terms, the moral orientation in task-oriented relations is characterized by the values of universalism, affective neutrality, and achievement. In contrast, the moral orientation of social-emotional relations is characterized by the values of particularism, affectivity, and ascription (Diesing 1962: 90). Obligations to other people in a social-emotional relationship are based on their particular relationship to oneself rather than on general principles: They are strongest when relations are close and weakest when relations are distant. In a task-oriented relation, one strives to detach oneself from the objects of one's actions and to treat them all as equal, separate, interchangeable entities; in a social-emotional relationship, one is the focal point of myriad relationships that one strives to maintain and extend, since action takes place only within relationships (Diesing 1962: 91). Ascription is the opposite of the achievement value: It means that one's actions and obligations toward people spring solely from their relationship to oneself rather than as a response to something they have done.

5.4.4 Formal Versus Informal

5.4.4.1 Cognitive Orientation

The basic element in the schema related to this dimension has to do with whether one expects the people involved in the social situation to let their activities, forms of relationship, demeanor, and the like be determined and regulated largely by social rules and conventions or whether one expects such people to have the freedom to make and break their own rules as suit their individual and collective inclinations. In a formal relationship, one expects that the latitude for deviation from conventional forms of behavior is small, and that when one violates the rules, others will react negatively and one will be embarrassed (if the violation is unwitting). Since the rules are *usually* well-known and well-articulated in a *formal* relationship, it is apt to be characterized by more predictability and less surprise than an informal one. Hostile rather than friendly relations, unequal rather than equal ones, and impersonal rather than formal ones are more likely to be regulated than informal.

5.4.4.2 Motivational Orientation

Formal social relations appear to be related to a cluster of psychological tendencies. Murray (1938: 200–204) has described various elements of this cluster: the *need for order,* subsuming conjunctivity, sameness, deliberateness, and placidity. Although Murray's emphasis is on the enduring character of these psychological predispositions, it seems likely that the psychological tendencies underlying the *bureaucratic personality* (Merton 1957) can be elicited by bureaucratic structures. These tendencies have been well described by Merton in his classic paper on bureaucratic structure and personality and amply characterized in the literature on the obsessive-compulsive personality. The emphasis here is on how 'formal' situations can temporarily induce in otherwise nonbureaucratic and nonobsessive personalities psychological predispositions to value order, regulation, predictability, sameness, lack of surprise, and the like.

Informal relations tend to be more open, more particularistic, more frank, more flexible, more emotional, and more personal than formal ones. They have a more relaxed, improvisational character in which quickly formed, intuitive, and impressionistic reactions to the specific other in the particular situation largely determine one's behavior. In an informal relation, one's motivational orientation is more directed toward persons, whereas in a formal relation, it is more directed toward rules and authority. Emotion and conflict is more apt to be openly expressed in informal relations and avoided in formal ones. The more enduring psychological predispositions that are characteristic of the so-called hysterical personality and the field dependent person, resemble the situationally induced motivational orientations to be found in informal relations.

5.4.4.3 Moral Orientation

In many respects, the moral orientations to task-oriented and formal relations are similar; this is also the case for social-emotional and informal relations. Formal relations go beyond the values of universalism and affective neutrality or impartiality to include a moral orientation to the rules and conventions that guide social relations. One has an obligation to respect them and to conform to them. One's obligation is to the form of the relationship rather than to its spirit. In contrast, in an informal relationship, one is morally oriented to the spirit rather than the form of the relationship. It is the relationship to which one is obligated rather than to the rules that are supposed to regulate it.

In the preceding pages, for brevity's sake, I have discussed the psychological orientations characterizing each of the four dimensions of interpersonal relations as though the dimensions existed in isolation from one another. Of course, in doing so, 1 have not adequately characterized the psychological orientations characterizing the different types of interpersonal relations: Each type reflects a combination of different dimensions. The psychological orientation associated, for example, with an intimate relation fuses the orientation connected with the particular positions on the cooperative, social-emotional, equal, informal, and intense dimensions. Here, the psychological orientations arising from the different dimensions of the relationship are all concordant with one another. A threat to an intimate relationship might arise from a discordance on any of the dimensions: for example, from a competitive orientation rather than a cooperative one ("I am more giving than you are"); from a task-oriented rather than a social-emotional one ("You don't accomplish enough"); from a dependent rather than an equal one ("1 need you to protect me and to take care of me"); or from a formal rather than informal one ("I get upset in a relationship unless I always know what is expected, unless it has no surprises, unless it is always orderly and predictable").

From our discussion of the correlations among the different dimensions in the first section of this chapter, it is evident that there is more or less discordance among the psychological orientations related to the different dimensions in the different types of social relations. Thus, the psychological orientation associated with cooperation is more concordant with the psychological orientations associated with equality, informality, and social-emotional activities than with the orientations associated with inequality, formality, and task-oriented activities. However, many cooperative relations are task-oriented and/or unequal and/or formal. Where there is discordance among the different dimensions characterizing a relationship, it seems likely that the relative weights or importance of the different dimensions in the given type of relationship will determine the relative weights of their associated psychological orientations. That is, if the task-oriented character of the relationship has stronger weight than the cooperative aspects, it will have more influence in determining the governing psychological orientation. It also seems likely that the more extreme the location on a given dimension, the more apt is that dimension to have the key role in determining the nature of the psychological orientation: In a situation that is extremely formal and only slightly cooperative, the psychological orientation will be determined more by the situation's formality than by its cooperativeness.

5.5 Some Relevant Research

In the opening paragraph of this chapter, I stated that the causal arrow connecting psychological orientations and types of interdependence is bidirectional: A psychological orientation can induce or be induced by a given type of interdependence. Here, I would go further and indicate that the cognitive, motivational, and moral components of a psychological orientation can each induce one another—hence, they are likely to be found together—and each of the components can induce or be induced by a given type of interdependence. The foregoing assumptions proliferate into a great number of testable, specific hypotheses that I do not have the space to elaborate in this chapter. To illustrate, however, these hypotheses would predict a two-way causal arrow between specific modes of thought and specific types of social relations. Thus, a 'bureaucratic' social situation will tend to induce "obsessive-compulsive" modes of thought and obsessive-compulsive modes of thought will tend to 'bureaucratize' a social relationship. They would also predict that a competitive social relationship will tend to increase the psychological weight or importance of the differences in values between oneself and one's competitors, whereas a cooperative relationship will tend to increase the psychological importance of the similarities in values between oneself and one's fellow cooperators. We would also hypothesize that a tendency to accentuate the differences in values between oneself and others is apt to induce a competitive relationship, whereas a tendency to accentuate the similarities is likely to induce a cooperative relationship. Further, it can be predicted that different principles of distributive justice will be associated with different types of social relations: A fraternal relationship will be connected with the principle of equality, a caring relationship with the principle of need, a hierarchical organization with the principle of equity, a power struggle with the principle of "winner-take-all." Each of these different principles can induce different modes of thought and different types of social relations when experimentally introduced into an otherwise unstructured social situation. For all of the various hypotheses that entail two-way causal arrows, from an experimental point of view, the independent variables are the ones that are manipulated by the experimenter and the dependent variables are the ones that are affected by the manipulated variables.

Some of the hypotheses suggested by the theoretical ideas presented in this chapter have been tested in my laboratory (Deutsch 1973) and by many other researchers working in a variety of areas in social psychology. However, many of these ideas have not yet been systematically investigated. Here, I wish to describe briefly two dissertation studies. One has recently been published (Judd 1978); data for the other are now being analyzed. Both were conducted in our laboratory and reflect our interests in the relation between types of interdependence and modes of thought.

In the first of these studies, Judd (1978) argues that competitive processes in attitude conflicts are characterized by a tendency to accentuate the evaluative differences between one's own position and the position of the person one is arguing with. One of the ways in which this might be done is by emphasizing those conceptual dimensions along which there are larger differences. Hence, Judd hypothesized that in a competitive attitude conflict, the conflicting parties will come to see their positions as being relatively dissimilar and this will be accomplished by heightening the evaluative centrality of those conceptual dimensions that best distinguish between the positions.

For cooperative processes, he argued that parties have the mutual goal of learning more about the issue under dispute. An emphasis on conceptual dimensions along which positions differ significantly may well lead to a more competitive conflict; therefore, Judd hypothesized that a cooperative orientation will motivate individuals to deemphasize those dimensions that best discriminate between the positions and to emphasize dimensions along which there is less of a difference. Thus, a cooperative orientation between conflicting parties will lead to the heightened perception of position similarity as a result of lowered evaluative centrality of the most discriminating dimensions and heightened evaluative centrality of less discriminating dimensions.

Judd came to the interesting conclusion that the perceptions of the similarity-dissimilarity of positions induced by one's orientation (competitive or cooperative) to a conflict will be mediated by *conceptual* changes in the way we look at the issue under dispute. We will come to place more evaluative emphasis on some dimensions and less on others, and these changes may be relatively long-lasting.

Judd's research was designed to test this hypothesis. Pairs of subjects were assigned positions on how national health insurance should be organized, an issue about which they did not have strong opinions. These positions differed along three attribute dimensions; positions of pairs being highly distant on one dimension, less distant on a second, and identical on the third. Distance positions along dimensions and content of dimensions were varied independently so that Judd's hypothesis could be tested independently of dimension content. Subjects were asked to either discuss or debate the issue under either a cooperative or competitive orientation. Following this, judgments of similarity of positions were gathered and dimensional evaluative centrality was measured in order to test the hypothesis under investigation.

The results of the experiment strongly confirmed its underlying hypothesis: Competition led to decreased perceived similarity between the positions, and the dimension on which positions differed most was most evaluatively central; cooperation had opposite effects. In other words, the competitive orientation led the competitors to develop conceptual structures, related to the issue under dispute, that accentuated the differences between them and made these differences more attitudinally significant to them; in contrast, the cooperative orientation led the disputants to develop conceptual structures that emphasized the similarities in their positions and made the similarities more emotionally important to them.

An experiment by William A. Wenck, now in progress in our laboratory, is also concerned with the relation between types of interdependence and modes of thought. In his study, Wenck is investigating the effects on modes of thought and types of social relations of three different distributive systems: (a) *winner-take-all,* where whoever contributes the most to the group receives the total outcome or reward received by the group; (b) *equity,* where the group's outcome is distributed to the individuals in proportion to their respective contributions to the group; and (c) *equality,* where the group's outcome is shared equally by all its members.

Wenck's investigation of the correlates of these three distributive systems derives from my (Deutsch 1976) characterization of them:

1. The *winner-take-all* system is associated with a 'macho,' power-oriented mentality; it also is associated with a high risk-taking, gambling orientation. This mode of thought is much more prevalent in men than in women. It is common in social conditions of disorder, intense competition, widespread illegality, violence, or poverty. It is common in frontier societies, in societies lacking a middle class, in illegal organizations, in adolescent male gangs, in warring groups, etc. It can be elicited by challenges to basic values, by unregulated competition, by an atmosphere of violence and illegality, by anything that stimulates greed or desperation.
2. The *equity* system is associated with an economic mode of thought that is characterized by quantification, measurement, calculation, comparison, evaluation, impersonality, and conversion of unique values to a common currency. It is a cool, detached, future-oriented, analytic, tough-minded mode of thought that appeals to universalistic values, logical reasoning, and objective reality rather than particularistic values, intuition, emotion, and subjective considerations. It is more prevalent in men than women. It is common in societies characterized by a stable hierarchical order, regulated competition, a developed economy,

technological advancement, and a large middle class who are neither poor nor rich. It is elicited by conditions that stress productivity, efficiency, objectivity, impersonality, detachment, individualism, and instrumentalism.
3. The *equality* system is associated with a particularistic, social-emotional orientation that is characterized by reliance on intuition, empathy, and personal feeling as a guide to reality. It is a holistic, involved, related, present-oriented 'soft' mode of thought in which the reality of others is apprehended from their inside rather than from the outside. Unlike the equity orientation, it is more prevalent in women than men. It is common in fraternal societies and in small cohesive groups that stress friendship, intimacy, loyalty, personal attachments, mutual respect, individual dignity, and cooperation. It is elicited by conditions that emphasize the bonds with others and the symmetrical-reciprocal character of these bonds.

Wenck's study employs a very involving, three-person task in which the group's outcome is determined by the activities of all three persons. The group outcome is distributed to the individuals according to one of the three distributive principles described earlier. The task permits a variety of individual behaviors: The participants can work independently, they can help one another, and they can harm one another. After working on the task for 30 min, the subjects are interrupted, and are administered a number of different instruments to obtain the dependent measures. Several questionnaires get at subjects' strategy in the task, their self-concepts as they worked on the task, their orientation toward other subjects, and their perception of others' orientation toward them as they all worked on the task. Adjective checklists elicit the motives and emotions that were activated during work on the task. In addition, projective techniques are employed to obtain the subjects' views of the group and of themselves as they worked in the group.

At this writing, the data have been collected and not yet completely analyzed. However, preliminary analyses show the following significant results: The 'equality' groups were more productive than the "equity groups" who were in turn more productive than the "winner-take-all" groups; the 'autobiographies' composed by the subjects for the roles they developed in the three experimental conditions differed from one another in expected ways; "winner-take-all" subjects characterized their thoughts and feelings as being more 'aggressive,' "risk-taking," 'ruthless,' 'selfish,' 'rougher,' 'unsharing,' and 'changeable' than did the subjects in the other two conditions; the subjects in the 'equality' condition described themselves as more 'nurturant,' 'affiliative' 'cooperative,' and 'altruistic' than did those in the other two conditions; self-characterization of the subjects in the 'equity' condition fell in between the "winner-take-all" and 'equality' conditions. It is apparent that the results that have been analyzed so far are in accord with the basic ideas underlying the experiment.

5.6 Concluding Comments

In this chapter, I have advanced several theses. First, different types of social relations can be characterized in terms of their positions on a number of basic dimensions of interpersonal relations. Second, each of the different types of social relations has associated with it distinctive psychological orientations. A psychological orientation is a complex consisting of interrelated cognitive, motivational, and moral orientations. Third, the causal arrow connecting psychological orientations and types of social relations is bidirectional: A psychological orientation can induce or be induced by a given type of social relationship. And, fourth, the various elements (cognitive, motivational, and moral) of a psychological orientation tend to be consistent with one another.

My argument is not that social relations determine psychological orientations without regard to the personalities of the individual participants, nor is it that psychological orientations induce distinctive social relations without regard to the nature of the social situation confronting them. My thesis is rather that there is a tendency for consistency between psychological orientations and social relations that will lead to change in one or both until congruence between the two has been largely achieved. In some circumstances, it will be easier to change psychological orientations; in others, social relations can be more readily altered. I have not addressed the problem of what determines how a conflict between one's psychological orientation to a relationship and the nature of that relationship will be resolved. This is an important problem for future work.

One final comment: My discussion throughout this paper has been of "ideal types" of social relations. Actual social relations are inevitably more complex than my discussion would suggest. An intimate love relationship, for example, is often characterized by considerable ambivalence: There are not only strong positive elements manifest in the relations but also intense anxieties latent within it; there are quarrels as well as embraces. In addition, it must be recognized that relationships develop and change. Apart from my brief discussion of Fig. 5.1, I have not attempted to characterize the dynamics of relationships. This, too, is an important problem for future work.

References

Abelson, R.P., 1975: "Representing Mundane Reality in Plans", in: Bobrow, D.G.; Collins, A.M. (Eds.): *Representation and Understanding* (New York: Academic Press).

Abelson, R.P., 1976: "Script Processing in Attitude Formation and Decision-Making", in: Carroll, J.S.; Payne, J.W. (Eds.), *Cognition and Social Behavior* (Hillsdale, NJ: Lawrence Erlbaum Associates).

Atkinson, J.W.; Feather, N.T. (Eds.), 1966: *A Theory of Achievement Motivation* (New York: John Wiley & Sons).

Bales, R.F., 1953: "Task Roles and Social Role in Problem-Solving Groups", in: Maccoby, E.E.; Newcomb T.M.; Hartley, E.L. (Eds.), *Reading in Social Psychology*, 3rd edition (New York: Henry Holt and Co.).
Banton, M., 1967: *Race Relations* (New York: Basic Books).
Bartlett, F.C., 1932: *Remembering* (Cambridge, England: Cambridge University Press).
Bateson, G., 1955: "A Theory of Play and Fantasy", in: *Psychiatric Research Reports,* 2: 39–51.
Benne, K.D.; Sheats, P., 1948: "Functional Roles and Group Members", in: *Journal of Social Issues*, 4: 41–49.
Cartwright, D.P.; Zander, A.F. (Eds.), 1968: *Group Dynamics: Research and Theory* (New York: Harper *and* Row).
Deutsch, M., 1949a: "A Theory of Cooperation and Competition", in: *Human Relations,* 2: 129–152.
Deutsch, M., 1949b: "An Experimental Study of the Effects of Cooperation and Competition upon Group Processes", in *Human Relations,* 2: 199–232.
Deutsch, M., 1962: "Cooperation and Trust: Some Theoretical Notes", in: Jones, M.R. (Ed.): *Nebraska Symposium on Motivation*, Volume X (Lincoln: University of Nebraska Press).
Deutsch, M., 1973: *The Resolution of Conflict: Constructive and Destructive Processes* (New Haven: Yale University Press).
Deutsch, M., 1979: "Education and Distributive Justice: Some Reflections on Grading Systems", in: *American Psychologist,* 34: 391–401.
Diesing, P., 1962: *Reason in Society* (Urbana, Illinois: University of Illinois Press).
Freud, A., 1937: *The Ego and the Mechanisms of Defense* (London: Hogarth Press).
Goftman, E.F., 1974: *Frame Analysis* (New York: Harper Arrow).
Head, Sir Henry, 1920: *Studies in Neurology* (Oxford: Oxford University Press).
Judd, C.M., 1978: "Cognitive Effects of Attitude and Conflict Resolution", in: *Journal of Conflict Resolution*, 22: 483–498.
Kelley, H.H.; Thibaut, J.W., 1978: *Interpersonal Relations: A Theory of Interdependence* (New York: John Wiley & Sons).
Kelley, H.H., 1979: *Personal Relationships: Their Structure and Processes* (Hillsdale, NJ: Lawrence Erlbaum Associates).
Lewin, K., 1951: *Field Theory in Social Science* (New York: Harper & Brothers).
Marwell, G.; Hage, J., 1970: "The Organization of Role Relationships: A Systematic Description", in: *American Sociological Review,* 35: 884–900.
McClelland, D.C.; Atkinson, J.W.; Clark, R.A.; Lowell, E.L., 1953: *The Achievement Motive* (New York: Appleton-Centuiy-Crofts).
Merton, R.K., 1957: *Social Theory and Social Structure*, Revised and enlarged edition (Glencoe, Illinois: Free Press).
Murray, H.A., 1938: *Explorations in Personality: A Clinical and Experimental Study of Fifty Men of College Age* (New York: Oxford University Press).
Neisser, U., 1976: *Cognition and Reality* (San Francisco: W. H. Freeman & Co.).
Rawls, J.A., 1972: *A Theory of Justice* (Cambridge: Harvard University Press).
Schank, R.C.; Abelson, R.P., 1977: *Scripts, Plans, Goals, and Understanding: An Inquiry into Human Knowledge Structures* (Hillsdale, NJ: Lawrence Erlbaum Associates).
Triandis, H.C., 1972: *The Analysis of Subjective Culture* (New York: Wiley Interscience).
Weber, M., 1957: "The Theory of Social and Economic Organization", in: Henderson, A.M.; Talcott, Parsons (Trans. and Eds.) (Glencoe, Illinois: Free Press).
Weiner, B., 1974: *Achievement Motivation and Attribution Theory* (Morristown, New Jersey: General Learning Press).
Wish, M.; Deutsch, M.; Kaplan, S.J., 1976: "Perceived Dimensions of Interpersonal Relations", in: *Journal of Personality and Social Psychology,* 33: 409–420.
Wish, M.; Kaplan, S.J., 1977: "Toward an Implicit Theory of Interpersonal Communication". in: *Sociometry*, 40: 234–246.

International Center for Cooperation and Conflict Resolution

THE MORTON DEUTSCH
INTERNATIONAL CENTER
FOR COOPERATION AND
CONFLICT RESOLUTION
TEACHERS COLLEGE | COLUMBIA UNIVERSITY

The ICCCR was founded at Teachers College, Columbia University, in 1986 under the direction of Professor Emeritus Morton Deutsch, Ph.D., one of the world's most respected scholars of conflict resolution. Professor Deutsch, an eminent social psychologist, has been widely honored for his scientific contributions involving research on cooperation and competition, social justice, group dynamics, and conflict resolution. He has published extensively and is well known for his pioneering studies in intergroup relations, social conformity, and the social psychology of justice. His books include: *Interracial Housing* (1951); *Theories in Social Psychology* (1965); *The Resolution of Conflict* (1973); *Distributive Justice* (1985); and *The Handbook of Conflict Resolution: Theory and Practice* (2000).

The Morton Deutsch International Center for Cooperation and Conflict Resolution (MD-ICCCR) is an innovative center committed to developing knowledge and practice to promote constructive conflict resolution, effective cooperation, and social justice. We partner with individuals, groups, organizations, and communities to create tools and environments through which conflicts can be resolved constructively and just and peaceful relationships can thrive. We work with sensitivity to cultural differences and emphasize the links between theory, research, and practice. While many conflict resolution centers provide training and consulting, our practice is rooted in our own original, leading-edge scholarship.

P.T. Coleman and M. Deutsch, *Morton Deutsch: A Pioneer in Developing Peace Psychology*, SpringerBriefs on Pioneers in Science and Practice 30,
DOI 10.1007/978-3-319-15440-4

Theory and Research

- Building on the theoretical legacies of Kurt Lewin and Morton Deutsch, we conduct basic and applied research on theory related to conflict, justice, cooperation, and systemic change.
- We work to bridge the gap between theory and practice in these areas.

Education

- We educate future leaders who will further the development of theory and practice in the interrelated areas of conflict resolution, cooperation, and social justice with the ultimate goal of understanding and supporting sustainable peace.
- We seek to increase public awareness of constructive methods for conflict prevention and resolution, of the many forms of oppression, and of strategies for overcoming social injustice in families, organizations, and communities worldwide and for fostering sustainable peace.

Practice

- We work with educational, non-profit, corporate, and governmental organizations to provide culturally sensitive and locally relevant services related to conflict, violence, justice, cooperation, and social change.
- We seek to broaden and enhance our international collaborative network.

Research Overview

Decades of research at the MD-ICCCR has addressed the question: *What determines whether conflicts move in a constructive or destructive direction?* While the answers to such questions are complex, we seek to identify the most fundamental factors that lead to qualitative differences in the dynamics of conflict and peace. Our research employs multiple disciplines, paradigms and methods to investigate the problems and opportunities of conflict in our world with the aim of fostering innovative practice and education.

This research has spawned new insights and new research questions, including:

- Are there optimal ratios of different motives that lead to constructive conflict?
- What determines fundamental differences in mediation strategies and the constructiveness of mediation?
- How do power differences between disputants affect conflicts and how can they be resolved constructively?
- How do cultural differences between disputants affect conflicts and how can they be resolved constructively?
- What determines whether conflicts over injustice and oppression move in a constructive or destructive direction?
- What are the fundamental dimensions of sustainable human development?
- Why do some types of conflicts seem impossible to resolve and what can we do to manage or resolve them?
- What determines the sustainability of peace?

Building on the foundational scholarship of Kurt Lewin and Morton Deutsch, the Center believes in the power of ideas to improve our world, and in the critical role leading-edge science plays in advancing and refining those ideas. Its approach is to develop conceptual models that address gaps in existing theory and research, often through eliciting insights from informed participants (local stakeholders and practitioners), and then to empirically test and develop the models using a variety of methods. Its scholarship bridges the theory-practice gap in our field by bringing new insights from research to bear on important technical and social problems, and by honoring practical expertise in the development of new theory. Work on such complex problems requires to integrate theory and research from a variety of different disciplines, to employ multiple methods such as case studies, surveys, lab experiments and computer modeling, and to work in multidisciplinary teams. The Center links its research to contemporary social problems, and communicate its findings to both scholarly audiences and the general public.

Education Overview

Situated at Teachers College, a top-ranked graduate school of education, the IC-CCR is recognized for educational excellence. The Center offers a wide range of courses for scholar-practitioners in the areas of cooperation, conflict resolution, dynamical systems, and social justice. It develops and provides state-of-the-art instruction, training, and professional development for students, practitioners, educators, and organizational leaders. It continues to generate additional opportunities for our external educational work with non-profit organizations, agencies, and communities nationally and internationally. The ICCCR is committed to building relationships with a variety of organizations to allow students to gain practical experience. It provides a bridge between the academic community and experienced practitioners as we support and encourage a reflective scholar-practitioner model.

Website: http://icccr.tc.columbia.edu/

About the Authors

Morton Deutsch (USA) is E.L. Thorndike Professor and director emeritus of the *Morton Deutsch International Center for Cooperation and Conflict Resolution* (MD-ICCCR) at Teachers College, Columbia University. He studied with Kurt Lewin at MIT's Research Center for Group Dynamics, where he obtained his Ph. D. in 1948. He is well-known for his pioneering studies in intergroup relations, cooperation-competition, conflict resolution, social conformity, and the social psychology of justice. His books include *Interracial Housing*, *Research Methods in Social Relations*, *Preventing World War III: Some Proposals*, *Theories in Social Psychology*, *The Resolution of Conflict*, *Applying Social Psychology*, and *Distributive Justice*.

His work has been widely honored by the Kurt Lewin Memorial Award, the G.W. Allport Prize, the Carl Hovl and Memorial Award, the AAAS Socio-Psychological Prize, APA's Distinguished Scientific Contribution Award, SESP's Distinguished Research Scientist Award, and the Nevitt Sanford Award. He is a William James Fellow of APS. He has also received lifetime achievement awards for his work on conflict management, cooperative learning, peace psychology, and applications of psychology to social issues. In addition, he has received the Teachers College Medal for his contributions to education, the Helsinki University medal for his contributions to psychology, and the doctorate of humane letters from the City University of New York. He has been president of the Society for the Psychological Study of Social Issues, the International Society of Political Psychology, the Eastern Psychological Association, the New York State Psychological Association, and several divisions of the American Psychological Association. It is not widely known, but after postdoctoral training, Deutsch received a certificate in psychoanalysis in 1958 and conducted a limited practice of psychoanalytic psychotherapy for more than twenty-five years.

P.T. Coleman and M. Deutsch, *Morton Deutsch: A Pioneer in Developing Peace Psychology*, SpringerBriefs on Pioneers in Science and Practice 30,
DOI 10.1007/978-3-319-15440-4

Address: Prof. Dr. Morton Deutsch, *Morton Deutsch International Center for Cooperation and Conflict Resolution.* Professor of Psychology and Education. Program in Social-Organizational Psychology, Department of Organization and Leadership, Teachers College, Columbia University, Box 53, 525 West 120th Street, New York, NY 10027, USA.

Website: http://icccr.tc.columbia.edu/.

E-mail: deutsch@tc.columbia.edu.

A website with additional information on Morton Deutsch, including links to videos and a selection of his major covers is at: http://afes-press-books.de/html/SpringerBriefs_PSP_MortonDeutsch.htm

Peter T. Coleman (USA) holds a Ph.D. in Social/Organizational Psychology from Columbia University; Professor of Psychology and Education at Columbia University with a joint-appointment at Teachers College and The Earth Institute and teaches courses in Conflict Resolution, Social Psychology, and Social Science Research; Director of the *Morton Deutsch International Center for Cooperation and Conflict Resolution* (MD-ICCCR) at Teachers College, Columbia University; Chair of Columbia University's *Advanced Consortium on Cooperation, Conflict, and Complexity* (AC4); research affiliate of the *International Center for Complexity and Conflict*; The Warsaw School for Social Psychology in Warsaw, Poland. He currently conducts research on optimality of motivational dynamics in conflict, power asymmetries and conflict, intractable conflict, multicultural conflict, justice and conflict, environmental conflict, mediation dynamics, and sustainable peace. In 2003, he became the first recipient of the Early Career Award from the American Psychological Association, Division 48: Society for the Study of Peace, Conflict, and Violence. He edited: *Handbook of Conflict Resolution: Theory and Practice* (2000, 2006, 2014); *The Five Percent: Finding Solutions to Seemingly Impossible Conflicts* (2011); *Conflict, Justice, and Interdependence: The Legacy of Morton Deutsch* (2011); *Psychological Components of Sustainable Peace* (2012); *Attracted to Conflict: Dynamic Foundations of Destructive Social Relations* (2013); *Making Conflict Work: Harnessing the Power of Disagreement* (2014). He has authored over 70 journal articles and chapters, is a member of the United Nation Mediation Support Unit's Academic Advisory Council, a founding board member of the Leymah Gbowee Peace Foundation USA, and a New York State certified mediator and experienced consultant.

Address: Prof. Dr. Peter T. Coleman, Director, *Morton Deutsch International Center for Cooperation and Conflict Resolution*; Professor of Psychology and Education. Program in Social-Organizational Psychology, Department of Organization and Leadership, Teachers College, Columbia University, Box 53, 525 West 120th Street, New York, NY 10027, USA.
Website: http://icccr.tc.columbia.edu/.
E-mail: coleman@exchange.tc.columbia.edu.

About the Book

Celebrating the 95th birthday of Morton Deutsch this book presents ten major texts of this much honored social psychologist on war and peace. After serving in the U.S. Air Force during World War II and being awarded two Distinguished Flying Cross medals, he worked as a psychologist for a more peaceful world. Influenced by Kurt Lewin, who believed that nothing was as practical as a good theory, Deutsch did theoretical work on such issues as cooperation-competition, conflict resolution, and social justice relating to war and peace issues. As President of the Society for the Study of Peace, Conflict, and Violence, the Society for the Psychological Study of Social Issues, and the International Society of Political Psychology, he helped to foster social science efforts to make for a more peaceful world. This second volume presents 10 major texts by Deutsch as a leading social science activist on issues of war and peace—writing papers, making speeches, participating in demonstrations.

Contents

Introduction by Morton Deutsch—The Prevention of World War III: A Psychological Perspective—A Psychological Basis for Peace—Undoing Nuclear Insanity—Negotiating the Non-Negotiable—On Producing Change an Adversary—Educating for a Peaceful World—Psychological Components of Sustainable Peace: An Introduction (with Peter T. Coleman)—William James: The First Peace Psychologist—Developing a Global Community: A Social Psychological Perspective (with Eric C. Marcus and Sarah Brazaitis)—Deutsch's Ten Commandments Regarding Conflict Resolution

A website with additional information on Morton Deutsch, including links to videos and a selection of his major covers is at: http://afes-press-books.de/html/SpringerBriefs_PSP_MortonDeutsch.htm.

P.T. Coleman and M. Deutsch, *Morton Deutsch: A Pioneer in Developing Peace Psychology*, SpringerBriefs on Pioneers in Science and Practice 30,
DOI 10.1007/978-3-319-15440-4

MIX
Papier aus verantwortungsvollen Quellen
Paper from responsible sources
FSC® C105338

If you have any concerns about our products,
you can contact us on
ProductSafety@springernature.com

In case Publisher is established outside the EU,
the EU authorized representative is:
Springer Nature Customer Service Center GmbH
Europaplatz 3, 69115 Heidelberg, Germany

Printed by Libri Plureos GmbH
in Hamburg, Germany